Joseph Dalton Hooker

An account of the botanical collections made in Kerguelen's Island during the Transit of Venus Expedition in the years 1874-75

Joseph Dalton Hooker

An account of the botanical collections made in Kerguelen's Island during the Transit of Venus Expedition in the years 1874-75

ISBN/EAN: 9783741141249

Manufactured in Europe, USA, Canada, Australia, Japa

Cover: Foto ©berggeist007 / pixelio.de

Manufactured and distributed by brebook publishing software
(www.brebook.com)

Joseph Dalton Hooker

An account of the botanical collections made in Kerguelen's Island during the Transit of Venus Expedition in the years 1874-75

BOTANY.

OBSERVATIONS ON THE BOTANY OF KERGUELEN ISLAND. *By J. D. Hooker, P.R.S.*

THE history of the botany of Kerguelen Island (also called Kerguelen's Land, and Desolation Island), previous to the visit of the Rev. Mr. Eaton, the last and most complete explorer of its flora, is a very brief one. It commences with the visit of Capt. Cook during his third voyage, in the narrative of which the vegetation of the island is thus described by Mr. Anderson, the surgeon of the "Resolution: " " Perhaps no " place hitherto discovered in either hemisphere, under the same parallel of latitude, " affords so scanty a field for the naturalist as this barren spot. The verdure which " appears, when at a little distance from the shore, would flatter one with the expec- " tation of meeting with some herbage; but in this we were much deceived. For " on landing we discovered that this lively colour was occasioned only by one small " plant, not much unlike some sorts of *Saxifrage*, which grows in large spreading " tufts, to a considerable way up the hills." Mr. Anderson proceeds then to give some particulars of this plant (*Azorella Selago*, Hk. f.), of the cabbage (*Pringlea antiscorbutica*, Br.), of two small plants found in boggy places, which were eaten as salad, one "almost like garden cress and very fiery" (probably *Ranunculus crassipes*, Hk. f.), the other very mild and "having not only male and female, but what bota- " nists call *androgynous* plants" (? *Callitriche*). He adds to these a coarse grass (*Poa Cookii*, Hk. f.), and a smaller sort which is rarer (probably *Deschampsia antarctica*, Hk.); a sort of goose-grass (? *Cotula plumosa*, Hk. f.), and another small plant much like it (this I do not recognise). "In short," he says, "the whole " catalogue of plants does not exceed 16 or 18, including some sorts of moss and a " beautiful Lichen " (*Neuropogon Taylori*, Hk. f.) "which grows higher upon the " rocks than the rest of the vegetable productions. Nor is there the least appear- " ance of a shrub in the whole country."

The date of Cook's visit was the summer of 1776, and the specimens obtained by Mr. Anderson were deposited in Sir Joseph Banks' Herbarium, which subsequently became the property of the nation, and is preserved in the British Museum. Not having been poisoned, all the Kerguelen Island plants were, when I examined them in 1843, much injured by insects, and many were entirely destroyed.

From 1776 till 1840, when the Antarctic Expedition under Capt. (afterwards Admiral Sir James) Ross, anchored in Christmas Harbour, Kerguelen Island is not known to have been visited by any ship of war, or by the Discovery or Surveying ships of any nation, though it had become the frequent resort of English and

American sealers. During the stay of the above-named expedition all the plants enumerated by Anderson as found by him in mid-summer were refound in mid winter, together with many more, amounting to nearly 150, of which 18 were flowering plants; the other large classes being mosses and Hepaticæ 35, Lichens 25 and Algæ 51. These have all been described in the botany of the voyage (Flora Antarctica, Part II., 1847).

The next visit of naturalists to Kerguelen's Land was that of the "Challenger" Expedition in January and February 1874, when Mr. Moseley collected most diligently both in Christmas Harbour and on the east coast 60 to 70 miles south-east of it He found 23 flowering plants in all, including three European weeds, all annual and doubtless imported by sealing parties (*Cerastium triviale*, *Poa pratensis* and *annua*), and three species not in the collections of the Antarctic Expedition (two *Ranunculi* and an *Uncinia*). He also procured flowering specimens of the two endemic genera *Pringlea* and *Lyallia*, and made large accessions to the cryptogamic flora, especially from the southern localities visited. Mr. Moseley had also the good fortune to land upon Marion Island, 1,650 miles to the west of Kerguelen Island and on Yong Island (of the Heard group), about 120 miles to the south-east of it neither of which had been previously visited by any naturalists, and in both of which he found some of the most peculiar of the Kerguelen plants.

Mr. Eaton arrived at Kerguelen Island with the Transit of Venus Expedition early in October 1874, and left towards the end of February 1875, during which time he collected diligently, chiefly at Royal Sound, Swains' Bay, and Observatory Bay. He obtained nearly all the flowering plants of previous explorers, and added very largely in the Cryptogams, especially to the Algæ.

Nearly contemporaneous with Mr. Eaton's visit was that of the American Transit Expedition, on which Dr. Kidder was the naturalist. He arrived in September 1874 and left in January of the following year, having explored some of the same localities as Mr. Eaton. His collections, amounting to about 90 species, are described in the bulletin of the U. S. National Museum, No. 3, issued in 1876 by the Government Printing Office of Washington. The flowering plants and ferns are revised by Prof. A. Gray; the mosses are described by Thos. P. James; the Lichens by Prof. E. Tuckerman, and the Algæ by Dr. W. G. Farlow. Except amongst the Lichens, there are very few novelties. Dr. Kidder adds a list of seven plants from the Crozets, all identical with Kerguelen Island species.*

The botanical results of the German Transit Expedition to Kerguelen Island are not yet published.

The three small archipelagos of Kerguelen Island (including the Heard Islands) Marion and Prince Edward's Islands, and the Crozets, are individually and collectively the most barren tracts on the Globe, whether in their own latitude or in

---

* He also mentions "a small vine with blue flowers growing amongst scoriæ," of which no specimens were collected. This is probably some endemic plant unknown to botanists.

any higher one, except such as lie within the Antarctic Circle itself; for no land even within the N. Polar area presents so impoverished a vegetation.

The chief interest attached to the flora of these archipelagos lies in the indication it affords of their being, in all probability, the remains of a much larger land area, which, though peopled with plants mainly from the southern extreme of S. America, 4,000 miles to the westward, possessed an endemic flora of its own, which included forest trees of considerable dimensions. Before, however, proceeding to discuss the relationships of their floras, I shall describe that of the largest and the only one that is at all well known.

As pointed out in the "Flora Antarctica," the prevalent features of the vegetation of this island as then known were Fuegian; one species of flowering plant alone, of those that are not peculiar to the island, being characteristic of any other flora, namely, the *Cotula plumosa*, which is found elsewhere only in the Auckland and Campbell Islands, south of New Zealand. More recent collections have confirmed and even strengthened this Fuegian affinity, for of the three additional flowering plants procured by subsequent explorers, one is Fuegian (*Ranunculus trullifolius*), another (*R. Moseleyi*) is closely allied to a Fuegian species, and the third one, *Uncinia compacta*, is a native of the mountains of New Zealand and Tasmania, and this is so nearly allied to a Fuegian species that it may prove to be a form of a plant common to all high southern latitudes.

Not only has a further knowledge of the Kerguelen Island flora strengthened its known affinities with the Fuegian, but recent discoveries in the latter flora have done so too; some of the Kerguelen's grasses especially proving to be more closely allied to Fuegian species than was suspected. The discovery of the flowers of the endemic Kerguelen genus *Lyallia* is another instance of this affinity. In the Flora Antarctica, judging from the fruit alone, the flowers being unknown, this remarkable plant was provisionally placed in *Portulaceæ*, its resemblance in habit and foliage to the andine genus *Pycnophyllum* being indicated. Complete specimens collected by Moseley prove its close relationship to the latter genus, in juxta-position with which it had indeed been placed in the Genera Plantarum, where both had been referred correctly to *Caryophylleæ*.

The elements of the Phænogamic flora of Kerguelen Island may be thus classified :—

1 Endemic genus, which has no near ally—*Pringlea antiscorbutica*.

1 Endemic genus allied to an Andean one—*Lyallia kerguelensis*.

6 Endemic species allied to American congeners—*Ranunculus crassipes* and *Moseleyi, Colobanthus kerguelensis, Acæna affinis, Poa Cookii, Festuca kerguelensis.*

5 species common to Fuegia but not found elsewhere: *Ranunculus trullifolius, Azorella Selago, Galium antarcticum, Festuca erecta, Deschampsia antarctica.*

6 species common to America, and also to New Zealand and the islands south of it. *Tillœa moschata, Montia fontana,* Callitriche obtusangula,* Limosella aquatica,* Juncus scheuzerioides, Agrostis Magellanica.* (Most of these are aquatic or marsh plants, and those marked with an asterisk are also European, and very widely dispersed.)

2 species found elsewhere but not in Fuegia, *Cotula plumosa,* common to Lord Auckland's group and Campbell's Island south of New Zealand, and *Uncinia compacta,* a native of the mountains of Tasmania and New Zealand.

This American affinity of the Kerguelen Island flora thus clearly established by its flowering plants is very strongly manifested by its Cryptogams, amongst which, however, the only evidence of migration from South Africa occurs. This is the case of *Polypodium vulgare,* a widely distributed fern in the north temperate zone, but known in the southern only from the Cape Colony, Marion, and Kerguelen Islands; what is further curious respecting it is, that the Kerguelen Island individuals are referable to a variety with pellucid veins, hitherto known only from the Sandwich Islands.

As to the local grouping of the Kerguelen Island plants, that of the Phœnogams is not altogether in harmony with the Cryptogams, the former seeming to be by far the most ubiquitously dispersed of the two groups.

All the plants hitherto collected have been from two areas, one, Christmas Harbour, in the extreme north, extending about five miles either way; the other, considerably larger, occupies the south-east coast, and following it extends for about 40 miles. The distance between these areas is about 60 miles in a N.W. and S.E. direction. Of the Phœnogamic plants, 19 were found in the northern area, nearly every one of which was also found in the south-eastern one, where but two additional species were collected; whereas of the 150 Cryptogams found in the northern area, a large proportion were not found in the south-eastern, where, however, nearly four times the number of species were obtained. Again, whilst but one fern was found in the north, four occur in the south-east. Of 35 *Musci* and *Hepaticœ* collected at Christmas Harbour by the Antarctic Expedition, hardly half were found at Swain's Bay, Betsy Cove, or Royal Sound, which localities yielded about 80 additional species. Nearly 50 marine Algæ were collected at Christmas Harbour, of which 18 did not occur in the south-eastern coasts, where upwards of 30 additional species were obtained. In the case of the *Lichens,* the discrepancy is still more marked, but this is possibly more apparent than real, and is to be attributed in part to the difficulty of defining the species and recognizing them from descriptions; and in part to the difficulties caused by the irreconcilable views of Lichenologists as to the limits of the species of this order.

Whatever other causes there may be for this anomalous distribution, one, no doubt, is the nature of the Christmas Harbour area. This is almost occupied by transverse valleys that run east and west completely across the north tip of the island, from sea to sea, are bounded by hills 1,200 feet high, and are perennially swept by terrific blasts from the westward. There are, hence, no shelter on land for the terrestrial flora, and no quiet bays for the proper development of a varied marine vegetation; facts which may very well account for the paucity of Cryptogams in Christmas Harbour, but not for the presence there of nearly all the flowering plants of the island. Turning again to the south-eastern area, its more sheltered valleys and land-locked harbours favour not only a greater development of Cryptogams, but also a far greater luxuriance of the Phænogams than obtains in Christmas Harbour; which last fact renders the absence of additional species of Phænogams to the south-eastward all the more remarkable.

The question remains, granting that the great majority of the Phænogams of Kerguelen Island are derived from South America, how was their transport effected? Though this question cannot be satisfactorily answered by a reference to the facilities for distant transport possessed by the fruiting organs of the Kerguelen Island plants, it is only proper to refer to these organs in some detail. Obviously, regarding the whole flora, the plants with the most minute seeds or spores and the water-plants are the most widely distributed. Under these categories come—1. The Fungi, of which all but 2 of the 8 species found are widely dispersed over the globe. 2. The marine Algæ, of which only 8 out of the 74 are peculiar to the island. 3. The fresh water Algæ, of which 28 out of 80 are regarded as endemic. 4. The aquatic and marsh Phænogams, 8 in all, of which 6 are widely dispersed.

Of the Phænogams, whether aquatic, marsh, or terrestial, none have appliances for wide dispersion except the hooked style of the *Ranunculus*, the reversed barbs of the *Acæna* (a most powerful aid), and the hooked organ attached to the fruit of *Uncinia*, also a very adequate aid. None of the others have any aid to dispersion, though they have small seeds or fruits.

Turning to the natural agents of dispersion, winds are no doubt the most powerful, and sufficient to account for the transport of the Cryptogamic spores; these, almost throughout the year, blow from Fuegia to Kerguelen Island, and in the opposite direction only for very short periods, but appear quite insufficient to transport seeds over 4,000 miles. Oceanic currents have, doubtless, brought the marine Algæ; but the transport of the seeds of the freshwater plants, of the grasses, and of the two plants with hooked and barbed appendages to the fruit, is not apparent in the case of a country that has no land birds but an endemic one (the Chionis), and of which the water birds come to land only or chiefly at the breeding season, and this after long periods of oceanic life in a most tempestuous ocean. Even supposing that the sea birds which habitually breed in Kerguelen Island did visit Fuegia between the periods of incubation, it is difficult to imagine that any seeds that had adhered to their beaks, feet, or bodies on leaving the latter country would not have been removed by the buffets of winds and waves over upwards of 4,000 miles of ocean.

The supposition that more land formerly existed along the parallels between

Fuegia and Kerguelen Island, possibly in the form of islands, remains as the forlorn
hope of the botanical geographer. By such stepping stones the land birds, so
numerous in the Falkland Islands (which lie in the direction of such hypothetical
islands), and of which the vegetation is identical with that of colder South America,
might, favoured by the prevalent westerly gales, have passed from thence to Kerguelen Island, having adhering to them fruits and seeds. The absence of such
birds from the present Avi-fauna of Kerguelen Island offers no obstacle to such a
speculation, as such immigrants would on arrival speedily be destroyed by the pre-
datory gull and petrels of the island.

Various phenomena, of very different relative value and nature, but common to
the three archipelagos, Kerguelen, the Crozets, and Marion, favour the supposition of
these all having been peopled with land plants from South America by means of
intermediate tracts of land that have now disappeared; in other words, that these
islands constitute the wrecks of either an ancient continent or an archipelago
which formerly extended further westwards, and that their present vegetation consists of the waifs and strays of a mainly Fuegian flora, together with a few survivals
of an endemic one.

The extreme southern point of South America, from lat. 52–54° and long. 70° W.
comprising Fuegia, is deflected to the eastward. Following its general direction, the
Falkland Islands group is the first land met with (in long. 60° W.); its vegetation is
comparatively rich and exclusively Fuegian; it has, no doubt, been brought mainly
by the land and freshwater birds which abound there, and are identical with
Fuegian ones. South Georgia is the next land met with to the eastward, in long.
35° W. and 54° S.; of its vegetation nothing is known except for the scanty observations recorded in Cook's voyage, which indicate its botanical identity with the
Fuegia.

Of Bouvet Island, the assumed position of which is long. 5° E. and 54° S., nothing
is known; it was searched for in vain by the Antarctic Expedition in 1843. Marion
Island is 37° E. and 46° S., and the Crozets, in 48° E. and 47° S., are respectively about
1,650 and 1,200 miles west of Kerguelen Island, and there is no land intermediate
between them. Now, from such specimens as have been obtained of the vegetation
of the first of these Islands by Mr. Moseley,* it appears to be almost identical
with that of Kerguelen Island; that is, to be Fuegian with the addition of some of
the peculiar Kerguelen Island types,† and the same remark applies to the Crozets,‡
facts from which Mr. Moseley has drawn identically the same conclusions as those
to which I had arrived thirty-five years previously from a consideration of the Kerguelen Island flora alone. He says, speaking of Marion Island (Linn. Journ. XV.,

* Journ. Linn. Soc. XIV., 387 and XV., 484.
† Marion Island contains several Fuegian species not hitherto found in Kerguelen Island, namely,
Ranunculus biternatus, Hymenophyllum tunbridgense, and probably a *Hierochloe* (the scented grass
mentioned by Moseley), together with a Cape fern *Aspidium mohrioides* and an *Asplenium.*
§ See Kidder in Bull. U.S. Nat. Mus. No. 3, p. 31.

85), "the occurrence of *Pringlea* on the island, as also on the Crozets and Kerguelen
Island, point to an ancient land connection between these islands, which the
antiquity and extent of denudation of the lavas would appear to bear out. It is
difficult to see how such seeds as those of *Pringlea* could have been transported
from one island to another by birds; and the seeds seem to be remarkably
perishable; besides the distinctness of the genus points to a former wide extension
of land on which its progenitors became developed. The existence of fossil tree
trunks in the Crozets and Kerguelen Island points to similar conditions.

In the Flora Antarctica, I say, p. 220, referring to the time required for the
formation of the innumerable superimposed beds of volcanic rocks, as observed by
me in Kerguelen's Land, and for the growths and destructions of successive forest
vegetations that once clothed the island, and are now imbedded in strata at great
depths, that this time is sufficient "for the destruction of a large body of land
" to the northward of it, of which St. Paul's and Amsterdam Island may be the
" only remains; or for the subsidence of a chain of mountains running east and
" west, of which Prince Edward's Island, Marion, and the Crozets, are the exposed
" peaks." And, at p. 240, when discussing the structural peculiarities of the
Pringlea, I say, "However loth we may be to concede to any of our vegetable pro-
" ductions an antiquity greater than another, or to this island (Kerguelen) a posi-
" tion to other lands wholly different from that it now presents, the most casual
" inspection of the land where this plant now grows will force one of the two
" following conclusions upon the mind, either that it was created after the extinc-
" tion of the now buried and for ever lost vegetation, or that it spread over the
" island from another and neighbouring region, where it was undisturbed during
" the devastation of this, but of whose existence no indication remains." *

It remains to indicate the faint traces of relationship which the Kerguelen Island
vegetation presents with those of a few other spots of land in a lower latitude, and
that might be supposed to share some of its peculiarities. Of these the nearest are
Amsterdam and St. Paul's Islands, the names of which are often transposed in our
best maps (even in the Admiralty South Polar Chart of 1839). They lie about 800

* These ideas, suggesting the hypothesis that the existing distribution of plants is dependent on former
geographical relations of land and sea, suggested themselves to me during my visit to Kerguelen Island in
1840. The first attempt to apply similar views in extenso to the conditions of a botanically well-known
country was in the late Professor Edward Forbes' paper "on the distribution of endemic plants, more espe-
" cially those of the British Islands, considered with regard to geological changes." "Brit. Assoc. Reports
" for 1845." It had, however, been previously enunciated by Lyell, who thus accounted for the identity of
the Sicilian animals and plants with those of the surrounding Mediterranean shores.

He supposes these to have "migrated from pre-existing lands, just as the plants and animals of the
" Phlægrean fields have colonised Monte Nuovo since that mountain was thrown up in the 16th century,"
and further on he says, "we are brought therefore to admit the curious result, that the flora and fauna of
" the Val di Noto, and some other mountain regions of Sicily, are of higher antiquity than the country
" itself, having not only flourished before the lands were raised from the deep, but even before they were
" deposited beneath the waters." Principles of Geology, Ed. v. iii., p. 444, &c.

miles to the N.E. of Kerguelen Island, in 78° E. long. ; the northernmost, Amsterdam
Island, is nearly on the 38th and St. Paul's on the 39th parallel of latitude, so they
both are very little south of the latitude of the Cape of Good Hope.

I have brought together, in a paper published in the Journal of the Linnæan
Society (vol. xiv. p. 474), all the little that was then known of the flora of these
islands, which, like Kerguelen, are volcanic.

Their scanty vegetation is on the whole more temperate than antarctic, and
approximates to that of S. Africa in containing such genera as *Phylica, Spartina,*
and *Danthonia.* Their fern flora is very interesting ; one fern only is common to
Kerguelen (*Lomaria alpina*), one (*Nephrodium antarcticum*) is peculiar, though
allied to a Mauritian species, and two others (*Blechnum australe* and *Asplenium
furcatum*) are natives of the Cape and other countries ; but what is most singular
is, that neither the *Polypodium vulgare* nor *Aspidium mohrioides* have been found in
either island, though the former is common to the Cape, Marion Island, and Kerguelen's Land, and the latter to the two first of these localities.

Tristan d'Acunha, in 12° W. long. and 37° S. lat., and the adjacent islets called
Nightingale and Inaccessible, all nearly in the latitude of Amsterdam Island and
the Cape of Good Hope, are the only other islands whose vegetation demands a
passing notice here.* Their flora is essentially Fuegian, with an admixture of Cape
genera, but with none of those characteristics of Kerguelen Island. Of Cape types,
it contains a *Pelargonium* and an abundance of both the *Phylica* and *Spartina* of
Amsterdam Island, together with species of *Oxalis* and *Hydrocotyle.* The Fuegian
and Falkland Island plants of Tristan d'Acunha and its islets, which have not hitherto
been found in the islands south and east of them, are however more numerous than
are the Cape genera even, and include *Cardamine hirsuta, Nertera depressa, Empe-
trum nigrum,* var. *rubrum, Lagenophora Commersoniana,* and *Apium australe ;* and
it contains besides the strictly American genus *Chevreulia.* Two land birds, both
peculiar, are common in the Tristan group, and they possess a water hen, which has
a representative in Africa and S. America. I am not aware whether land birds
are found in Amsterdam Island ; if so, they may help to account for the wonderful
fact of the Tristan d'Acunha *Phylica* and *Spartina* being found also in it, though
separated by 3,000 miles of ocean.

In conclusion, I have to state that no trace of the mountain flora of S. Africa
has been found in any of the southern groups of islands.

---

* For the latest account of this group see Moseley in Journ. Linn. Soc. XIV., 377.

ENUMERATION OF THE PLANTS HITHERTO COLLECTED IN KERGUELEN ISLAND BY THE "ANTARCTIC," "CHALLENGER," AND "BRITISH TRANSIT OF VENUS" EXPEDITIONS.

I.—*Flowering Plants, Ferns, Lycopodiaceæ, and Characeæ.*
By J. D. HOOKER, P.R.S.

1. **Ranunculus crassipes**, *Hook. f. Fl. Antarct.* 224, t. 81.

Christmas Harbour, Observatory and Swain's Bay, Royal Sound (a form with petioles 5-7 inches long).

I have nothing to add to what I have said of this species in the Antarctic Flora, beyond that I can hardly doubt its being a derivative form of the Fuegian *R. biternatus*, Sm., with which it agrees in habit and its thick-walled beaked carpels, but differs chiefly in its robustness and simple leaves. *R. biternatus* has been found by Moseley in Marion Island, where it presents every character of the American plant.

2. **Ranunculus trullifolius**, *Hook. f. Fl. Antarct.* 226, t. 82 A.

In streamlets and lakes, Royal Sound, Swain's Bay, Betsy Cove; *Moseley, Eaton, Kidder.* (Fuegia and the Falklands).

Glaberrimus, caulibus prostratis radicantibus. *Folio* longe crasse petiolata, obovato-oblonga trulliformia v. fere orbicularia, apice obtuse 3 5-dentata v. lobata, carnosula, nervis obscuris ; auriculis petiolaribus membranaceo-dilatatis. *Flores* ad nodos solitarii, brevissime pedicellati. *Sepala* 3, orbicularia, concava, membranacea. *Petala* 3, sepalis æquilonga, obovato-oblonga v. spathulata, 3-nervia, nervo medio medium versus fossa nectarifera instructo. *Stamina* pauca. *Carpella* numerosa ; matura cuneiformia, compressa, dorso incrassata, stylo gracili subulato.

I described this species in the Flora Antarctica from very imperfect specimens gathered by myself in the Falklands in mid-winter, along with the very similar *R. hydrophilus*, Gaud., and from a careful examination of the remains of the only flower found, which resembled in petals, sepals, and stamens those of its neighbour, I supposed it to be closely allied to it. Good specimens gathered by Cunningham in the Straits of Magalhaens, and by Eaton in Kerguelen, prove that it belongs to another section of the genus, differing from *R. hydrophilus* in the usually trimerous perianth and the long style of the flattened ripe carpels. *R. trullifolius* is, in fact, referrable to St. Hilaire's genus *Casalia* (now reduced to *Ranunculus*), and its nearest ally is *R. bonariensis*, Poiret (*R. Kunthii Trian.* and *Planch.*), which differs by its ovate crenate leaves, long-peduncled flowers, and absence of style in the ripe carpels. *R. hydrophilus*, again, is probably a form of *R. adscendens*, St. Hil. (*R. humilis*,

** B

Collie, in Hook. Bot. Beech. Voy. p. 4, t. ii.), which has similar minute subglobose ripe carpels without a style.

R. monanthos, Philippi of Chili, and *R. hemignostus* Steud. of Peru, are probably forms of *R. trullifolius*, which, as our figure shows, is a very variable plant in foliage and structure.

The *Ranunculus*, sp. 3, not in flower, of Kidder (Bull. U. S. Nat. Mus. 3, 21), of which Gray says it can hardly be a form of *trullifolius*, no doubt is this, if, as I apprehend, the term *caudate* as applied to the leaves is a misprint for *cordate*.

PLATE I., Figs. 1–5.—Plants in different states; of *natural size*; 6, 7, reduced leaves and stipules; 8, sepal; 9, petal; 10 and 11, stamen; 12, immature, and 13, mature carpels :—all *enlarged*.

3. Ranunculus Moseleyi, *Hook. f.*; pusillus, glaberrimus, acaulis, foliis radicalibus, petiolo in laminam obovatam v. oblongam integerrimam dilatato, floribus solitariis pedunculatis minutis 3–4-meris, petalis lineari-obovatis obtusis eglandulosis, staminibus 4–7, carpellis 10–12 maturis oblique subglobosis in stylum brevem gracilem abrupte attenuatis.—Ranunculus an nov. sp.; *Oliver, in Journ. Linn. Soc.* XIV., 389.

In the lake at Christmas Harbour, *Moseley*.

A very diminutive species, resembling in size and habit *R. limoselloides*, Muell, of Australia, but differing in the carpels, &c. In the latter respect it more nearly approaches *R. crassipes*, from which it differs in all other respects. Its allies are, no doubt, to be found amongst the S. American water-loving species.

PLATE II., Fig. 1—1 and 2, plants of *natural size*; 3, leaf; 4, flower; 5, sepal; 6, petal; 7, stamen; 8, immature; and 9, mature carpel:—all *enlarged*.

4. Pringlea antiscorbutica, *Br. MSS.; Fl. Antarct.* 238, t. 90, 91; *Kidder in Bull. U. S. Nat. Mus., No.* 321; *Oliver in Journ. Linn. Soc.* XIV., 389; *Dyer in Proc. Linn. Soc.* 1874, xxxiv.; *Hook. f.* l. c.

Throughout the island.—(Marion, Crozets, and Heard Islands).

Sepala lineari-oblonga, obtusa, membranacea, pilosa. *Petala* 0 in exemplaribus perplurimis a nobis scrutatis, in paucis 1–4, unguiculata, apice roseo-tincta, inconspicua, caduca. *Stamina* 6, subæqualia, filamentis elongatis complanatis, 4 longioribus per paria sepalis anticis posticisque opposita; antheræ magnæ, lineari-oblongæ, virescentes; pollen sphericum. *Disci* glandulæ 0 v. valde inconspicuæ. *Ovarium* oblongum, hirsutum, 2-loculare, carpellis lateralibus; stylus brevis, validus, glaber, stigmate capitato obscure 2-lobo dense villoso.

In the Proceedings of the Linnæan Society 1874, p. xxxiv, I have indicated the evidence of *Pringlea* being a wind-fertilized member of a natural order most or all the species of which are insect-fertilized. These indications are the usual absence of petals and disk-glands, the exserted anthers and long-tufted papillæ of the stigma to which is to be added the absence of winged insects in Kerguelen Island. In reference to the last statement, it is a curious fact that wingless flies abound in the island, and on this very plant. Moseley, Journ. Linn. Soc. xv., 54, in his notes on Kerguelen botany, mentions an apterous fly as big as a blow-fly, nestling at the base of the leaves of *Pringlea* and laying its eggs in the fluid which is caught there; every cabbage yielding ten or a dozen specimens. He adds that he did not observe whether it climbs to the inflorescence in sunny weather.

Mr. A. W. Bennett, Proc. Linn. Soc. 1874, xxxix., has described the pollen of *Pringlea* as differing from that of nearly all other Crucifers in being much smaller and perfectly spherical, instead of ellipsoid with three furrows. This he considers to be a striking confirmation of my suggestion that the plant is wind-fertilized, and which is further confirmed by the total absence of hairs on the style.

Moseley found one plant with 28 flower-stalks, three of the one season growth, the others appearing to belong to eight preceding seasons.

It is a remarkable fact that all attempts to grow this plant in England, Scotland, and Ireland have failed; the young plants, after attaining a height of a few inches and a good crown of leaves, have invariably succumbed to the combined effects of summer's heat, and the attacks of the common parasite fungus, *Cystopus candidus*, which infests the *Capsella Bursa-pastoris*. Some few, out of many hundreds, sown at different seasons and under very varied conditions, survived one winter, but perished in the following summer.

PLATE II., Fig. 3.—1, 2, 3, apetalous flowers; 4, monopetalous, and 5, tripetalous flowers; 6, petal; 7, ovary; 8, the same laid open; 9, ovule :—all *enlarged*.

5. **Colobanthus kerguelensis,** *Hook. f. Fl. Antarct.* 249, t. 92.

Christmas Harbour, Swain's Bay, &c. (Heard Island, *Moseley*.)

(*Stellaria media L.*)
Introduced by sealers.

(*Cerastium triviate*, Link.)
Introduced by sealers.

6. **Lyallia kerguelensis,** *Hook. f. Fl. Antarct.* 548, t. 122; *Kidder in Bull. U.S. Nat. Mus., No. 3.,* 22. *Oliver in Journ. Linn. Soc. XIV.* 390. *Dyer in Proc. Linn. Soc.* 1874, xxxiv.

Christmas Harbour and Royal Sound.

The flowers have been described from Kidder's specimens by Asa Gray, and from Moseley's by Oliver and Dyer, the descriptions agreeing well. The stamens, which appear to be almost constantly three and hypogynous, are stated by Oliver to be variable in position. Kidder retains it in *Portulaceæ*, but Bentham and I had long previously placed it in *Caryophylleæ* in the Genera Plantarum and next to *Pycnophyllum*, a position which the discovery of the flowers confirms. It has many of the characters of *Colobanthus*, especially the andrœcium.

PLATE II., Fig. 2.—1, plant, of *natural size;* 2, leaves; 3, flower and bract; 4, flower laid open; 5, stamen :—all *enlarged*.

7. **Montia fontana**, *L.*

Common in wet places. (Marion Island, *Moseley*, and widely distributed in the N. and S. temperate regions).

8. **Acæna affinis**, *Hook. f. Fl. Antarct.* 268, t. 96 B.

Common throughout the island. (Marion and the Crozet Islands).

Called Kerguelen's tea, and used as a febrifuge by whalers (Kidder).

Unlike the *Pringlea* and *Cotula*, this plant has grown and flowered at Kew from roots sent by Moseley.

9. **Callitriche verna**, *L.*; Subsp. *obtusangula*. C. obtusangula, *Le Gall. Hegelm. Monog. Gall. Callit.* 54. C. antarctica, *Engelm. ex Hegelm.* l. c.; *Kidder in Bull. U.S. Nat. Mus., No.* 3, 23. C. verna, *Hook. f. Fl. Antarct.* 272.

Common in wet places. (Marion and Heard Islands, *Moseley*, and widely distributed in the N. and S. temperate regions).

From a drawing of the ripe fruit which I made when in Kerguelen in 1840, I have no hesitation in referring this to Subspecies *obtusangula*, as Hegelmeyer did from his examination of my dried specimens. The fruit lobes are nearly semi-circular and each pair is united by about two thirds of their faces. The free portions are obtusely trigonous at the back. Two forms are common in Kerguelen, as elsewhere in the south temperate zone, one aquatic with long stem and proportionally large spathulate leaves, the other smaller, terrestrial, suberect, with obovate or oblong leaves; this flowers the most abundantly.

10. **Tillæa moschata**, *D.C.* Bulliarda moschata, *D'Urv.*

Abundant in moist places near the sea. (Marion Island, *Moseley*, Crozets, *Kidder*; widely spread in high southern latitudes).

11. **Azorella Selago**, *Hook. f. Fl. Antarct.* 284, t. 99.

Very abundant throughout the island. (Marion and Heard Islands, Moseley; Crozets, Kidder; Fuegia; Mac Quarrie Island.)

Kidder remarks that the flowers are greenish yellow, not pale pink as I found them to be in winter. Also, that the leaves have not the bristles on the faces of the lobes as figured in the Flora Antarctica. I find them on specimens from all localities.

Moseley observes, in reference to this plant at Marion Island, that the mounds it forms evidently retain and store up a considerable amount of sun's heat, and that this fact probably explains its peculiar mode and form of growth, and that of many otherwise widely different Antarctic plants. He found that a thermometer plunged into the heart of a hummock rose to 50°, when the temperature of the air was 45°.

12. **Galium antarcticum**, *Hook f. Fl. Antarct.* 303 bis.

Common, but not found at Christmas Harbour. (Crozets, *Kidder*; Fuegia and Falkland Islands.)

Kidder remarks that the flowers are distinctly pedicelled, and as often 4- as 3-merous, and even 5-merous ones occur. Eaton's specimens confirm this.

13. **Cotula** (Leptinella) **plumosa,** *Hook. f. Fl. Antarct.* 26 and 308, t. 20.

On cliffs, especially near the sea, often forming immense luxuriant blue-green atches where the soil is enriched by the dung of birds and seals. (Crozets, *Kidder*; ord Auckland, Campbell's, and Mac Quarrie Islands.)

Reputed by the whalers to be a prompt and effectual emetic. Through a typographical omission of the word *not* at p. 308 of the Antarctic Flora, this plant is tated to be found on the continent of America. The genus *Leptinella* is reduced a *Cotula* in the Genera Plantarum. This plant, like the *Pringlea*, proved so upatient of heat in this country, that of innumerable seedlings raised at Kew to veral inches high all perished.

14. **Limosella aquatica,** *L.*

Common in the freshwater lagoon at Christmas Harbour. (Fuegia and all emperate regions.)

A very small form, with the leaf-blade hardly broader than the petiole. *Stamens* ncluded. *Ovary* globose; style rather long.

15. **Juncus scheuzerioides,** *Gaud.; Hook. f. Flor. Antarct.,* 79, 358.

Common in spongy places. (Fuegia, the Falkland, Lord Auckland, and Campbell's Islands.)

16. **Uncinia compacta,** *Br.; Boott in Hook. f. Fl. Tasman,* ii. 103, t. 153 B.

Royal Sound and Observatory Bay, *Moseley, Eaton.* (Mountains of Tasmania nd New Zealand.)

17. **Deschampsia antarctica,** *Hook. Ic. Pl. t.* 150 (Aira); *Hook. f. Fl. Antarct.* 377, *t.* 133.

Common and ascending to considerable altitudes. (Fuegia, Falkland Islands, South Shetlands.)

A true *Deschampsia,* as that genus is now defined, by its 4-toothed flowering glume and free caryopsis, *Munro.*

18. **Agrostis magellanica,** *Lamk.; Hook. f. Fl. Antarct.* 373. A. antarctica, *ibid.* 373, *t.* 132. A multicaulis, *ibid.* 95.

Common throughout the island. (Marion and Heard Islands, *Moseley;* Chili, Fuegia, Falkland, and Campbell's Islands.)

Since the publication of this plant as *A. antarctica,* I have examined a specimen of Lamarck's *A. magellanica* named by Nees in Arnott's Herbarium, and find it to be identical. Further, Munro informs me that it is fairly described by Trinius in his "Agrostideæ," and by Kunth in his supplemental volume (p. 175) from a Lamarckian specimen; he adds that the Kerguelen specimens agree with these descriptions, except in the flowering glume being larger and much longer than the ovary. This glume is sometimes obtuse or rounded, at others deeply divided. The beard on the callus, which is very indistinct on the Kerguelen's plant, is conspicuous on some Fuegian ones.

19. **Poa Cookii**, *Hook. f.* ; *Fl. Antarct.* 382, t. 139 (Festuca).

Forma 1.; foliis culmum superantibus, panicula elongata interrupta.

Forma 2.; foliis culmum superantibus v. æquantibus acuminatis pungentibus panicula densa sub-cylindracea.

Forma 3.; foliis culmum æquantibus subacutis v. obtusis, panicula minore laxiore, spiculis paucifloris coloratis.

Abundant and ascending to a considerable height :—Forma 1. Christmas Harbour ; Forma 3. Royal Sound, on a high hill, *Eaton*. (Marion and Heard Islands, *Moseley*).

This fine grass should, unquestionably, be referred to *Poa* (as now defined by the compressed flowering glume, &c.), along with its near congener *Dactylis cæspitosa* * of Fuegia and the Falklands, from which it differs, amongst other characters, in never forming tussocks. It is scarcely specifically distinct from *P. foliosa*, Hook. f. Handbook of N. Z. Flora 338 (Festuca foliosa, *Fl. Antarct.* i. 99, t. 55 ; *Fl. Nov. Zeald* i. 308) ; and this, again, from the Fuegian *Poa lanigera*, *Nees* (Festuca fuegiana, *Fl Antarct.* 380). The flowering glumes are often obscurely, or not at all toothed. The spikelets are 3–5-flowered and $\frac{1}{4}$–$\frac{1}{3}$ in. long (not eight lines as misprinted for three lines in the Antarctic Flora). A. Gray remarks of Kidder's specimens that they seem to be male only.

Poa pratensis, L.

Introduced by sealers.

Poa annua, L.

Introduced by sealers.

20. **Festuca erecta**, *D'Urv.*

Common and ascending to a considerable elevation. (Fuegia and the Falkland Islands.)

Often forming tussocks ; panicles green or purplish.

21. **Festuca kerguelensis**, *Hook. f.* Triodia kerguelensis, *Fl. Antarct.* 379, *t.* 138 (*Poa*).

Common and ascending to 2,000 feet.

Spikelets sometimes 1-flowered. A very variable grass in stature, evidently allied to *F. erecta*, and more nearly still to *F. scoparia* (Fl. Antarct. 98 ; Fl. Nov. Zeald. i. 308), of which possibly it is a dwarf form, as suggested in the Handbook of the New Zealand Flora, p. 341. The naked base of the flowering glume, however, will always distinguish all the specimens I have examined.

Filices.

1. **Cystopteris fragilis**, *Bernh.*

Crevices of rocks near the hill-tops, Royal Sound, *Kidder*, *Eaton*. (Fuegia, Falklands, and N. and S. temperate regions generally.)

---

* The name *Poa cæspitosa* being occupied by Forster, though it is doubtful to what species it applies, I propose that of *flabellata* for the Tussock grass, which is the *Festuca flabellata*, Lamk.

2. **Lomaria alpina,** *Spreng.*

Common, often forming large beds, but not found at Christmas Harbour.
Marion Island, *Moseley;* Crozets, *Kidder;* all the colder S. temperate regions.)

3. **Polypodium** (GRAMMITIS) **australe,** *Mett.*

Crevices of rocks, Observatory Bay, *Kidder, Eaton.* (Marion Island, Moseley;
Fuegia, and all the colder S. temperate regions.)

4. **Polypodium vulgare,** L. *var.* Eatoni, *Baker,* venis pellucidis.

Crevices of rocks by running streams, Observatory Bay, *Kidder, Eaton.* (Marion
Island, Moseley; S. Africa; Sandwich Islands, and N. temperate hemisphere.)
This pellucid-nerved variety only occurs elsewhere in the Sandwich Islands.

Lycopodiaceæ.

5. **Lycopodum clavatum,** *L., var.* magellanicum; *Hook. f. Fl. Antarct.,*
13. L. magellanicum, *Swartz.*

Not uncommon throughout the island, but not met with at Christmas Harbour.
(Var. *magellanicum,* Marion Island, Moseley; Fuegia, and all the colder S. temperate
regions. The typical *L. clavatum* inhabits all northern cold damp climates.

6. **Lycopodium Selago,** *L.* var. Saururus, *Hook. f. Fl. Antarct.* 394.
L. Saururus, Lamk.

Not uncommon throughout the island. (Var. *Saururus,* Marion Island, *Moseley;*
Tristan d'Acunha, St. Helena, Bourbon, Peru. The typical form inhabits all damp
cold climates.)

Characeæ.

7. **Nitella antarctica,** *Braun.* N. Hookeri, *Reinsch in Journ. Linn. Soc.*
xv. 219. Chara flexilis, *Linn;* Fl. Antarct. 395.

In the Lake at Christmas Harbour; and in that next but one to the Observatory,
in Observatory Bay, *Eaton.*

II.—*Musci.*

By WILLIAM MITTEN, A.L.S.

The first investigation of the mosses of Kerguelen was made by Dr. J. D. Hooker during the voyage of the " Erebus " and " Terror " in the winter of 1840.

From the collections made by him there were described 31 species and varieties, which were arranged as 25 species in 11 genera. Of the whole number six species were considered to be new and undescribed, and the remainder to have been found in other regions. The most remarkable species contained in this collection are the *Schistidium marginatum, Weissia stricta,* and *W. tortifolia.*

During the visit of the Challenger, there were collected by Mr. Moseley, in the summer of 1874, 28 species, of which number 20 were additional to those discovered by Dr. Hooker. Sufficient materials were obtained to establish the presence of eight more genera, all previously known to occur in austral lands, four of the species appearing to be new. Twenty-eight species were obtained by Dr. Kidder of the American Transit Expedition, of which number 12 were additions to the Flora, two being described as new. Following the above come the collections made by the Rev. A. E. Eaton, pending the observations of the transit of Venus, which include 38 species, of which 17 were additional to the Flora of Kerguelen Island, three being undescribed, and by this collection three genera were also added ; thus raising the whole number of the species of mosses inhabiting the Island to 74. This, considering how much has been added by each collector to those which were previously known, is probably a low estimate of the entire moss flora.

No genera peculiar to Kerguelen are observable in the collections, unless a species here referred to *Blindia* and the *Schistidium marginatum* (here placed in *Streptopogon*) should be so considered. The remaining genera are universal in boreal as well as austral regions, with the exception of the three species of *Dicranum* all which belong to extra-European sections of that genus. Twenty-three of the Kerguelen mosses are considered identical with species found in the north of Europe and America, of these *Bryum alpinum* and *Brachythecium salebrosum* had not before been identified in the southern hemisphere.

A few distinct and well-marked species have been gathered in Kerguelen Island which are also found at great elevations on the Andes of Quito and of New Grenada. Of these *Mielechhoferia campylocarpa* and *Psilopilum trichodon* are conspicuous instances ; they probably inhabit the whole Andine chain. *Bartramia appressa, Brachythecium paradoxum,* and *Tortula Princeps* are found also in New Zealand and Tasmania ; but with the exception of *Dicranum kerguelense* there is no species which points to any connexion with the mosses of South Africa.

1. **Ditrichium australe**, *Mitt.* l. c. (Cynontodium). Lophiodon strictus, *Hook. f. et Wils. Fl. Antarct.* 130, t. LIX., Fig. 1.

In dense fulvous tufts, with old capsules, *Moseley.* (Lord Auckland's and Campbell's Islands.)

In all the specimens referred to this species the dry young foliage is fulvous, the older brown or black; the terminal leaves are frequently longitudinally twisted, otherwise their direction is the same as when wet; the lower portion of the leaf is in outline of an elliptic oblong figure, from which the nerve is continued in a straight line, and is rather suddenly carried out so as to appear without a margin of leaf; a transverse section shows it to be concave above and convex beneath; the apex is abrupt, rounded, and nearly flat, so as to appear as if dilated, and, as stated in the Flora Antarctica, the species is distinguished from most of its allies by this particular. The substance of the base of the leaf is composed of elongated cells which, although shorter towards the top of the dilated portion, are not dense, so that the entire expansion is of a pellucid fulvous colour, the nerve being everywhere smooth, with a few small teeth at its apex.*

2. **Ditrichium Hookeri**, *C. Muller Syn. I.*, 450 (Leptotrichum).

Royal Sound, with old capsules and young setæ, *Eaton.*

3. **Ditrichium conicum**, *Mont. in Ann. Sc. Nat. Ser.* 3, iv. 100. (Aschistodon.)

Near Vulcan Cave, barren, *Eaton.*

The imbrication of the leaves at the apices of the stems, when dry, so as to form an erect or curved point, renders this species not difficult to recognise in a barren state.

1. **Asiothecium vaginatum**, *Hook. Musc. Exot.* t. 141 (Dicranum).

* In the Journal of the Linnean Society, Sept. 1859, there was confused with the *Leptotrichum australe*, therein mentioned, the following apparently distinct species,—*D. punctulatum*, Mitt.; dioicum? dense cæspitosum, dichotome ramosum, folia inter se remotiuscula a basi erecta amplexante oblonga cellulis inferioribus elongatis superioribus abbreviatis rotundatis obscuriusculis veluti punctatis, subito in subulam patentem inferne canaliculatam apice angustam planiusculam denticulatam minutissime scabridam sublævem cellulis punctulatis areolatam producta, perichætii alia basi latiora et longiora parte subulato patentiora, theca in pedunculo breviusculo rubro parva ovali-cylindracea erecta leptoderma fulvo-fusca. Flos masculus in ramis terminalis, ovatus, e basibus foliorum dilatatis apice retusis vaginantibus involucratus. Distichium capillaceum, Fl. N. Zealand, II., 73.

Hab.—New Zealand, *Dr. Lyall.* Great Barrier Island, *Hutton and Kirk.* Fagus Forests, *Hopkins, Dr. Haast.*

In size colour and general appearance very similar to *D. australe*, having also the same, but narrower, flattened apices to its leaves; in the recurvation of the subulate portion from the top of the erect base it resembles *D. capillaceum*, and for this species Dr. Lyall's barren specimens were mistaken, although the leaves are not distichous, but so disposed that each fifth leaf occupies the same vertical position on the stem as the first counted from; the outline of the dilated base is not oval-elliptic as in *D. australe*, but oblong obtuse. The fruit in an old state is present on Dr. Haast's specimens; accompanying these fertile stems were many conspicuous male flowers, which do not appear to arise from the lower parts of fertile stems, but seem to be really distinct male plants.

**

Swain's Bay, *Eaton.*

Small barren stems, but not different from specimens from the Bogotian and Quitenian Andes.

1. **Blindia gracillima,** Mitt.   Dioica, laxe cæspitosa.   Caulis elongatus, gracillimus, inferne nudus, superne nudus, superne foliis remotiusculis laxe obtectus.   Folia anguste lanceoloto-subulata, pagina folii e cellulis angustis elongatis parietibus pellucidis usque ad $\frac{2}{3}$ nervi apice vix denticulati longitudinis anguste continuata; cellulis alaribus in auriculam parvam dispositis rubris; folia perichætialia erecta, basi obovata, convoluta, sensim subulato-attenuata, nervo longius excurrente. Theca in seta brevi flexuosa arcuata pendula, subresupinata, globosa; operculo oblique rostrato; peristomii dentibus rubris latis teneris integris vel rarius pertusis intus lævibus extus parietibus transversalibus prominentibus appendiculatis; annulo nullo; calyptra parva, viridis, nigrescens.   B. curviseta, *Mitt. in Linn. Soc. Journ.* XV., 193.

Royal Sound, in lakes, with young and nearly ripe fruit, *Eaton.*

Stems 2–4 inches long, forming loose tufts, the upper portions red, the lower black, denuded of leaves, and forming a loose entangled mass.   Leaves at the apices of the stems fulvous and shining, the lower all blackened; in their direction the upper leaves are but little changed when wet or dry; they are $1\frac{1}{2}$–2 lines long; the areolation consists of elongate cells separated by pellucid walls; at the angles of the base of the leaf the alary cells are distinct and red.   The nerve becomes indistinguishable at four-fifths of the whole length of the leaf, and is thence continued, and ends without forming a pungent point; leaves of the perichætium longer, and their dilated bases about twice the width of the cauline leaves.   Seta 2–2½ lines, straight in its lower half, thence to the capsule twisted and variously curved.   Capsule erect when dry, when wet with a swan's-neck-like curve, and so bent as to become horizontal; when mature spherical without any neck where it is affixed to the seta; colour reddish brown; substance thin but firm.   Operculum always obliquely beaked, at length of the same colour as the capsule.   Peristome perfectly formed; teeth red, broad at the base, thence with an even outline narrowed to their points, with the exception of a rare perforation there is no trace of their being composed of a double row of cells; at the base of the teeth the transverse divisions are close together, but above this they are much wider, and on turning the tooth on edge it is seen that each dissepiment of the articulations is prominent on the outer side, but not on the inner.   Spores small, round.   Calyptra coriaceous, brownish-green, deeply cleft, with a spreading base.

Tab. III., Fig. 1, plant of natural size; 2, cauline leaf; 3, perichætium with capsule; 4, portion of peristome; all *magnified.*

2. **Blindia microcarpa,** *Mitt. in Journ. Linn. Soc.,* XV., 65.   Monoica, pulvinatim cæspitosa.   Caulis dichotomus, fastigiatim ramosus.   Folia patentia, stricta, plus minus falcata curvatave, dimidio inferiore lanceolato superiore carinato

anguste attenuato, integerrima, nervo angusto percursa, cellulis elongatis alaribus in auriculam parvam fuscam dispositis areolata; perichætialia brevia, parva, ovata, convoluta, in acumen subulatum producta. Theca in pedunculo gracili foliis caulinis dimidio breviore erecta, parva, ovalis ; operculo subulato obliquo demum ore dilatato cyathiformi fusca; peristomii dentibus teneris ; calyptra parva, dimidiata. Flos masculus foliis propriis perichætialibus similibus inclusus.

Kerguelen Island, *Moseley*.

This is the species mentioned in the Flora Antarctica, p. 128, under *Weissia contecta*, as being present in the Hookerian Herbarium, its habitat unknown.

In compact, but not coherent tufts. Stems fastigiately branched, about an inch high. Foliage shining, but little altered in direction wet or dry. The minute capsule is scarcely conspicuous amongst the leaves. Leaves at the tops of the stems yellowish green, below brown, erect or slightly falcate, about $2\frac{1}{2}$ lines long, composed of elongate cells with pellucid walls; nerve pale brown and with the pagina gradually attenuated into a very narrow flat entire point ; alary cells at the angles of the base distinct, brown, forming sub-quadrate masses. Leaves of the perichætium $\frac{1}{4}$-$\frac{1}{2}$ as long as those of the stem, and quite concealed amongst them. Seta about 1 line long, straight, pale brown. Capsule as it reaches maturity appearing to pass from oval to nearly globular; after the fall of the operculum by the dilatation of its mouth it becomes cyathiform, with no distinct neck. Operculum with a very oblique subulate beak which is longer than the capsule. Peristome-teeth very thin, broad at base, narrowed upwards into entire points; transverse articulations remote. Calyptra small, coriaceous, brownish, scarcely reaching the base of the operculum. Male inflorescence in a small bud below the base of the perichætium.

Tab. III., Fig. ii. : 1, plant of natural size ; 2, cauline leaf ; 3, perichætium with comal leaf, capsule, and male flower ; 4, old capsule ; 5, portion of peristome ; all *magnified*.

3. **Blindia contecta**, *Hook. f. & Wils. Flor. Antarct.* 404 t. 58, f. 3. (Weissia).

Christmas Harbour, on rocks, barren, *Hooker*.

In this species the perichætium is composed of enlarged leaves as in *Stylostegium*, Schimp., but the capsule which is immersed has a peristome.*

---

* These three species afford some considerations respecting their mode of fructification. The genus *Blindia*, Bruch et Schimp., at first included only the European *B. acuta*, with the "perichætium vaginans distinctum," the perichætial leaves being as large as the cauline and dilated below. To this was added by C. Müller (in the Synopsis) *B. cæspiticia*, which had been made into the genus *Stylostegium* in the Bryologia Europea, differing from *Blindia* in its gymnostomous capsule immersed in enlarged but not vaginant perichætial leaves, in these particulars analogous to some species of *Grimmia* of the section *Schistidium*, in which *B. cæspiticia* had itself sometimes been placed. The distinction between *Blindia* and *Stylostegium* is reduced by the presence of a peristome in *B. contecta* (which may be said to be a *Stylostegium* with a peristome) by the immersed capsule in *Stylostegium*, and the exserted one in *Blindia*. In *B.*

c 2

1. **Dicranum** (Isocarpus, *Mitt.*) **tortifolium**, *Hook. f. et Wils. Fl. An-
arct.* 404, *t.* 152, *f.* 5 (Weissia).

Hab., Christmas Harbour, on gravelly banks, *Hooker.* Under the shoot of
waterfall near Vulcan Cove, with old capsules and young setæ, *Eaton.*
In compact tufts 1–1½ inch high. Foliage very green above, below becoming
brown. Old capsules black and shining ; young calyptras orange brown.

2. **Dicranum** (Isocarpus) **strictum**, *Hook. f. et Wils. Fl. Antarct.*, 40
l. 152, *f.* 4 (Weissia).

Christmas Harbour, on rocks near the sea, *Hooker.*

This has been described as dioicous, but the male flower is terminal on
branch arising some distance below the perichætium. The peristome has rathe
broad thin teeth ; in the solitary example which could be examined, the teeth ap
peared to be partly adherent in pairs, the median line is obsolete. This specie
is closely related to *D. tortifolium.*

3. **Dicranum** (Hemicampylus, *Mitt.*) **robustum**, *Hook. f. et Wils. F
Antarct.* 406, *t.* 152, *f.* 8, *var.* lucidum; D. pungens, *var.* lucidum, *Hook. f. e
Wils. t. c.*

Hab. Christmas Harbour, *Hooker, Moseley.*
Known only in a barren state.

3. **Dicranum** (Hemicampylus) **kerguelense**, *C. Müller, Syn. i.* 370. 1
Boryanum, *Schwaegr; Hook. f. et Wils. Fl. Antarct.* 406. D. dichotomum, *Beau
Prodr.* 51 (Cecalyphum).

Christmas Harbour, *Hooker.* On an elevated moor, Royal Sound, *Eaton.*

microcarpa the perichætium is formed of leaves reduced in size like those which usually include the anth
ridia, and thus another modification of the perichætium is produced, all other particulars being as in *Blind*
proper. Thus, by the difference in the leaves of the perichætium, the species are separable into sever
groups :—

Stylostegium, B. & S. ;—theca in perichætio e foliis caulinis ampliatis immersa.
Blindia, B. & S. ;—theca e perichætio e foliis basi vaginantibus caulinorum magnitudinis exserta.
Homogastrium ;—theca e perichætio microphyllo exserta.

The differences in the leaves of the perichætium between *Stylostegium* and *Blindia* are analogous
those which exist between the *Grimmiæ* of the sections *Schistidium* and *Grimmia;* between *Hedwigidiu*
and *Brannia;* between some *Bartramiæ* of the section *Leucomela* and *Eubartramia;* and also between th
Schlotheimiæ of the sections *Stegotheca* and *Euschlotheimia.* States of the perichætium analogous to th
observable in *B. microcarpa* occur chiefly in mosses which produce their fruit from the side of the ste
as *Anœctangium,* and in some species of *Fissidens.* Amongst the Neckeroid mosses perichætia may
observed in otherwise closely resembling species which are analogous to all three of the states here left
Blindia. Much time and many words might be saved in the description of mosses in which the perichætiu
is an important character, if some term at once conveying the essential part of the above information wer
employed, thus :—

Chanogastriati ;—perichætium e foliis elongatis ampliatisque hians=*Stylostegium,* Schistidium (Grim
miæ), *Hedwigia, Hedwigidium, Cryphæa, Neckera.*

Heterogastriati ;—perichætium e foliis elongatis inferne convolutis clausum=*Blindia, Dicranu*
Hypnum, &c.

Homogastriati ;—perichætium e foliis abbreviatis iis perigonii similibus formatum=*Blindia microcarp*
Anœctangium, Pyrrhobryum, &c.

In extensive tufts, with stems from 3–4 inches high, and fulvous green foliage, becoming when older, brown.

So far as can be seen from the small specimen in the Hookerian Herbarium of *Cecalyphum dichotomum*, it appears to be the same as the Kerguelen moss, as it was considered by Mr. Wilson. The chief distinction ascribed to *D. kerguelense* is to have the nerve vanishing towards the narrow flat point, and not as in *D. dicholomum* to have the nerve continued into the point and dentate on the back.

1. **Campylopus cavifolius,** *Mitt. Musc. Austr. Amer.* 87.

Kerguelen Island, in dense tufts, barren, *Moseley.* By some error this was enumerated in the Linn. Soc. Journal as *C. appressifolius.*

1. **Ceratodon purpureus,** *Linn. Sp. Pl.* 1575 (Mnium).

Hab.—Royal Sound and near Swain's Bay, in a dark purple almost blackened state, all barren, *Eaton.* (Heard Island, *Moseley.*)

This moss appears to be as common throughout the southern regions as it is in the northern. The southern states have generally a more robust appearance, but when *C. brasiliensis*, Hampe, from Brazil, *C. crassinervis*, Lorentz, from Valdivia, *C. capensis*, Schimp., from the Cape of Good Hope, and *C. convolutus*, Reichardt, from New Zealand, are compared side by side, the conclusion seems irresistible, that they are all forms of one species.

1. **Grimmia** (SCHISTIDIUM) **apocarpa,** *Linn. Sp. Pl.* 1579 (Bryum).

Christmas Harbour, *Hooker.* Cat Island, Royal Sound, *Eaton.*

A very small state; all the specimens unlike European, but not appearing to be really different.

2. **Grimmia** (SCHISTIDIUM) **falcata,** *Hook. f. et Wils. Fl. Antarct.* 101, *t.* 151, *f.* 8.

Christmas Harbour, on rocks and stones near a waterfall, *Hooker.*

This is either an aquatic species or an aquatic form of a species of which the form corresponding to rupestral states of *G. apocarpa* is unknown.

3. **Grimmia insularis,** *Mitt. in Journ. Linn. Soc.* XV., 73.

Heard Island, *Moseley.*

4. **Grimmia** (EUGRIMMIA) **Kidderi,** *James in Bull. U. S. Nat. Mus.*, 3, p. 25.

Near Swain's Bay, *Eaton.*

In small dense black cushions. Stems 3–4 lines high, with a few branches near the base, made up of repeated innovations from the base of the male flower, consisting of closely set widely ovate leaves, without diaphanous points, including a few antheridia. Leaves very small, canaliculate, margins erect, terminated by a short, nearly smooth hyaline point.

This ambiguous moss may be conjectured to represent a species near to the European *G. montana.*

5. **Grimmia** (DRYPTODON) **chlorocarpa,** *Brid., Mill. in Hook. f. Handb. New Zealand Fl.*, II., 426 (sub Rhacomitrium crispulum).

Kerguelen Island, *Moseley*.   Hill N.W. of Mount Crozier, barren, *Eaton*.
Very closely related to *G. Symphyodon* and *G. emersa*, C. Müller, and also to
D. crispulus, Hook. f. et Wils.; all are possibly forms of one species.

6. **Grimmia** (DRYPTODON) **crispulus**, *Hook. f. et Wils. Flor. Antarcl.* 124,
et 402, *t.* 57, *f.* 9.

Christmas Harbour, in gravelly beds of rivulets, *Hooker*.

7. **Grimmia** (RHACOMITRIUM) **lanuginosa**, *Dill.; Brid.* i. 215.

Hab.—Kerguelen Island, *Moseley;* Royal Sound and near Vulcan Cove, barren,
Eaton.

All the specimens are less robust than those collected by Dr. Hooker in Hermite
Island; from the whitening of the tips of the leaves they are very hoary.

Many of the specimens brought from southern regions which appear to differ in
only slight particulars from northern states have been described as distinct; of
these, *Rhacomitrium firmum* De Notaris, from Chili, is a fulvous brown moss,
R. Gerontioum, C. Müller (Hedwigia, 1870), is possibly the same.   *R. senile*,
Schimp. (Lechler, 1089, from Magellan), with leaf points crisped and hoary, *R.
incanum*, C. Müller (Hedwigia, 1870), from Cape of Good Hope, is, if specimens
from the top of Table Mountain belong to it, scarcely in any particular different
from Arctic examples.

8. **Grimmia** (RHACOMITRIUM) **protensum**, *A. Braun; Hook. f. et Wils.
Flor. Antarct.* 402.

Christmas Island, barren, *Hooker*.

9. [*G.* FRONDOSA, *James in Bull. U. S. Nat. Mus.* 3, 25, is another Kerguelen
Island species, found by Kidder.]

1. **Orthotrichum crassifolium**, *Hook. f. et Wils. Fl. Antarct.*, *p.* 125,
tab. lvii. *f.* 8.

Christmas Harbour, common, *Hooker*, *Moseley;* Royal Sound, *Eaton*.

The specimens from Kerguelen have the points of the perichætial leaves reaching to three-fourths of the length of the capsule, which is thus only emergent, and
in this respect they agree with some of the specimens gathered in Hermite Island
by Dr. Hooker.   No importance can be attached to this particular character, as in
Dr. Hooker's specimens from Lord Auckland's Islands, emergent and exserted
capsules may be seen on the same stems.

The capsules are either smooth or with a few folds regularly placed on one side,
the remainder being smooth, and are more urcolate than any of the specimens
collected by Dr. Hooker.

The inflorescence consists, as usual in the genus, of a male flower near the base
of the perichætium in all the specimens.

2. **Orthotrichum atratum**, *Mitt. in Linn. Soc. Journ., XV., p.* 66.   Monoicum.   Caulis humilis, cæspitosus.   Folia patentia, sicca incurva, laxe contorta, lanceolata, apice lata obtusiusecule acuta, nervo sub summo apice evanescente, cellulis

fere ubique parvis rotundatis obscuris ; perichætialia majora. Theca in pedunculo longitudine perichætii subæquali ovalis, lævis, sicca infra os contracta, inferne collo crasso; operculo convexo, rostro angusto; peristomii dentibus 16, vel plus minus cohærentibus 8. Calyptra nigro-fusca, calva, ad medium usque thecæ descendens, nitida.

Kerguelen Island, *Moseley.*

Stems not more than half an inch high. Leaves a line long; a few of the youngest greenish, the rest all black, coriaceous. Capsule pale straw-coloured, somewhat fleshy, smooth when deoperculate, very slightly contracted just below the mouth at the base, when dry shortly plicate.

In all its parts larger than *O. crassifolium*, with leaves twice as wide, and without the horny appearance; it is, however, more nearly allied to that species than to any other, and approaches in some respects the *O. anomalum*, Hedw., which ascends far towards the Polar regions.

3. **Orthotrichum rupestre,** *Schleich. ; Brid.* i. 279.

Royal Sound, with fruit nearly mature, *Eaton.*

The specimen is in good state, and appears to agree in all respects with the European, except that no internal peristome has been found; it does not correspond so well with either of the very closely allied species, *O. Sturmii* or *O. cupulatum,* which agree in being destitute of cilia.

1. **Zygodon Brownii,** *Schwaegr.* t. 317 *b.*

Kerguelen Island, *Moseley.*

The minute scrap rather establishes the fact that a species of the genus inhabits Kerguelen Island than provides materials for identifying with certainty that to which it is here referred.

Tortula (Syntrichia) **Princeps,** *De Notaris ;* Barbula Mülleri, *Bruch et Schimp., Bryol. Europ.* t. 28. T. Fuegiana, *Mitt., Journ. of Linn. Soc.,* Sept. 1859. *Musc. Austr. Amer.* 174. Barbula, S. magellanica, *C. Müller* in *Bot. Zeit.* 1862, 349; B. antarctica, *Hampe ;* Tortula antarctica, T. cuspidata, *et* T. rubella. *Hook. f. et Wils. Fl. Tasmanica, pl.* clxxii., *f.* 8, 9, 10.

Royal Sound, with abundant mature capsules; Observatory Bay, with older fruit, *Eaton.*

The first examination of the Kerguelen specimens yielded no male inflorescence, they were therefore considered to be *T. fuegiana*, with which in size, colour, and appearance they appeared to be identical, this being supposed to be a dioicous species, as no male flowers were observed in Lechler's Magellan specimens No. 1088, from Cabo Negro. The same specimens were again described by C. Müller as dioicous, under the name of *Barbula S. magellanica.* In seeking for the male flowers amongst Mr. Eaton's abundant specimens, it was, after the examination of many stems, ascertained that although no antheridia were present in the fertile flowers, a small proportion of the stems had a male flower without archegonia, either terminal on a short branch, or lateral from the growth of innovations. Finally it was discovered that there might be present on the same stem, flowers containing antheridia accompanied by others containing archegonia, and above both these another flower in which both organs were intermixed. Thus, with specimens in small quantity to examine, the inflorescence might be described as monoicous dioicous or synoicous, as might chance to happen to the investigator.

The European *T. Princeps* was at first correctly described by De Notaris as polygamous in the Bryologia Europea, where it is figured as *Barbula Mülleri*. It is there described as hermaphrodite, with a remark in a subsequent note that it occasionally produced flowers containing archegonia only. In Schimper's Synopsis and in the Bryologia Britannica it is simply stated to be synoicous. An examination of De Notaris's original specimen shows synoicous fertile flowers with innovations of the stem terminated by flowers with archegonia alone; in this particular coinciding with British specimens.

The distribution of this species appears to be very wide, and it would seem to be the preponderating if not the only species of the genus in southern regions. From N.W. America it extends to Mexico, Chili, and the Straits of Magellan; in Africa it is found at the Cape of Good Hope, and may be identical with the *Barbula mollis*, Schimp., of the Abyssinian Mountains; it occurs in N.W. India; it inhabits also New Zealand, Tasmania, and Australia, from whence several species have been described as dioicous, viz., *Barbula Latrobeana*, C. Müller (Bot. Zeit. 1864, 358), *B. Preissiana* (ejusd. Synops. I. 642), *B. pandurœfolia* (ejusd. et Hampe, Linnæa 1853, 493). No specimen, however, amongst those sent by Baron F. von Mueller to the Kew Herbarium has been examined without finding its inflorescence monoicous or synoicous. There is also *Tortula S. pusilla*, J. Angstr. from Magellan, described as dioicous? and *Barbula Lechleri*, C. Müller (Bot. Zeit. 1859, 229), as monoicous. All these species or supposed species may be well distinguishable, but if the certainty of the condition of their inflorescence is removed from their descriptions, the remainder becomes applicable to *T. Princeps*, in which the outline of the leaves even on the same stems is, as in the European *T. ruralis*, subject to a great amount of variation.

2. **Tortula** (Barbula) **serrulata**, *Hook. et Grev. in Brewst. Edinb. Journ.* i. 291, t. 12.

Kerguelen Island; a few small barren stems with other mosses, *Moseley*.

3. **Tortula** (Barbula) **erubescens**, *Mitt. in Hook. f. Handbook of New Zeald. Flora*, ii. 421 (Didymodon).

Kerguelen Island; a few fragments, *Moseley*.

Very closely related to the *T. rubella* so widely distributed in northern regions, differing chiefly in the longer operculum and larger size of the whole plant.

1. **Streptopogon australis**, *Mitt. in Linn. Soc. Journ.* xv. 66. Humilis.

Folia inferiora patentia, spathulato-ligulata, obtusiuscule acuta, nervo in apice desinente, margine apicem versus denticulata ; superiora duplo latiora, a basi erectiore sensim recurva, patentia, apice cum nervo in acumen longitudine variabile sensim educto, margine superne serrulata.

Royal Sound ; a single stem, *Ealon*. Two small stems amongst other mosses without precise locality, *Moseley*.

The small quantity found of this moss would be insufficient to give any idea of what might be supposed to be the usual appearance of the species were it not evidently a close congener to a very ambiguous moss found on thatch in the south of Britain, and which has been known first as a supposed gemmiferous variety of *Leplodontium flexifolium* (Sm.), and since as *Didymodon gemmascens*, Mitt. MSS. From this the Kerguelen species differs in the form of its lower leaves. In the British moss all the leaves are acuminate and tipped with a globular mass of individually obovate green gemmæ of a loose cellular substance, and gemmæ of the same form are present on the points of some of the upper leaves of *S. australis*.

Both species appear to be small, the British one is seldom more than half an inch in height ; the entire plant, excepting a few rootlets, and the rarely present archegonia, which are red, is of a yellowish green. In the dry state it affords nothing to attract observation, but when wet, every leaf being terminated by its mass of gemmæ, it is unlike any other European moss, excepting the more robust *Orthotrichum phyllanthum* (Brid.). It comes nearer to some species of *Streptopogon* ; the areolation of the leaves of *Calymperes* or of *Syrrhopodon* are widely different. The genus *Streptopogon* founded on *S. erythrodontus* (Tayl.), with the additional species discovered in the Quitenian Andes by Dr. Spruce, and those from the Bogotian Andes by Lindig and Weir, contains a number of species all seeming to have a tufted Orthotrichoid habit. They differ among themselves considerably, some of the Andean species having the leaf with a callous margin which is wanting in others, and the capsule immersed or shortly exserted from perichætial leaves which are not very different from the cauline. *S. mnioides*, Schw. t. 310 (*Barbula*), however, has the perichætium leaves much elongated, and different from those of the stem, simulating those of *Holomitrium*, and on this account should stand apart from the other species, thus—

STREPTOPOGON, *Wils.* Theca in perichætio e foliis caulinis subsimilibus immersa, emergens, vel breviter exserta. Calyptra breviter multifida.

CALYPTOPOGON, *Mitt.* Theca in perichætio e foliis elongatis a caulinis difformibus exserta. Calyptra profunde plurifida.

The first group contains all the species of which the fruit is known, and which correspond to the typical *S. erythrodontus*, together with probably some others which are known only in a barren state, including the two ambiguous species *S. australis* and *S. gemmascens*.

The second group consists of *S. mnioides* alone.

2. **Streptopogon? marginatus ;**—Schistidium marginatum, *Hook. f. and Wils. Flor. Antarcl.*, 399, *l.* 151. f. vi.

**   D

Christmas Harbour, forming large patches ou wet rocks, *Hooker*.

This, which appears destitute of peristome, is in other respects more nearly related to *Streptopogon* than to any other genus, and if included in it would occupy a position annalogous to that of *Stylostegium cæspiticium* and *S. conicetum* before mentioned under *Blindia*.

1. **Entosthodon laxus**, *Hook. f. et Wils. Fl. Antarct.*, 390, t. 151, f. 5. (Physcomitrium).

Christmas Harbour, *Hooker*. Royal Sound, barren, and Swain's Bay, with nearly mature capsules, *Eaton*.

Traces of an internal peristome are present within the external teeth.

1. **Bartramia** (PHILONOTIS) **appressa**, *Hook. f. et Wils. Ft. New Zealdd.* ii. 89, t. 86, f. 5.

Royal Sound, barren; Observatory Bay, with a few nearly ripened capsules; and hill N. W. of Mount Crozier, a tall barren slender state, *Eaton*.

2. **Bartramia** (PHILONOTIS) **australis**, *Mitt. in Hook. Handb. New Zeald. Flor.*, 448.

Swain's Bay and Royal Sound, all barren, *Eaton*.

The few small stems growing among other mosses appear to belong to this species.

3. **Bartramia** (BREUTELIA) **pendula**, *Hook. Muse. Exot.* t. 21.

Kerguelen Island, *Moseley*. Royal Sound; hill N.W. of Mt. Crozier; near Vulcan Cove, with abundant immature fruit, *Eaton*.

4. **Bartramia** (EUBARTRAMIA) **patens**, *Brid. Sp. Musc.* iii. 82.

Kerguelen Island, *Moseley*. Royal Sound, with old fruit; and hill N.W. of Mt. Crozier, *Eaton*.

5. **Bartramia** (EUBARTRAMIA) **robusta**, *Hook. f. et Wils. Ft. Antarct.* t. 59.

Kerguelen Island, *Moseley*. Royal Sound, with old capsules and young setæ rising, very fine tall specimens, and Swain's Bay, *Eaton*. (Heard Island, *Moseley*.)

[B. FLAVICANS, *Mitt.*, is enumerated by James as amongst the U. S. collections, collected at the rear of the American Transit House.]

1. **Bryum** (WEBERA) **nutans**, *Schreb.; Hedw. Musc. Frond.* i. t. 4.

Near Vulcan Cove; hill N.W. of Mt. Crozier, a small state with unripe fruit growing amongst *Psitopilum trichodon*, *Eaton*.

2. **Bryum** (WEBERA) **elongatum**, *Dicks*.

Swain's Bay, a single stem with ripe capsule, *Eaton*.

3. **Bryum** (WEBERA) **crudum**, *Hedw. Musc. Frond.* i. t. 88 (Mnium).

Kerguelen Island, *Moseley*. Swain's Bay, with fruit just mature, *Eaton*.

4. **Bryum** (WEBERA) **albicans**, *Wahlenb*.

Christmas Harbour, *Hooker, Moseley*. Near Vulcan Cove, *Eaton*. Specimens all barren.

5. **Bryum** (ECCREMOTHECIUM)**pendulum**, *Hornsch*.

Royal Sound; and Cat Island, Royal Sound, *Eaton*.

The inflorescence, which is usually synoicous in capsuliferous flowers, is sometimes accompanied by unisexual flowers upon the same stem.

6. **Bryum** (ECCREMOTHECIUM) **Eatoni,** *Mitt. in Journ. Linn. Soc.*, xv., p. 195. Synoicum. Caulis humilis, gracilis, innovationibus infra comalibus paucis ramosus. Folia erecto-patentia, inferiora minora, superiora elliptico-lanceolata, nervo in acumen tenue læve vel denticulis paucis asperum excurrente, margine limbo tenui e seriebus cellularum elongatarum 4–5 composito anguste reflexo integerrima, cellulis angustis limitibus teneris areolata ; folia comalia longiora, basi subauriculato-dilatata, angulis rotundatis laxis areolatis. Seta elongata, recta, apice anguste curvata. Theca pendula, sporangio ovato collo subæquilongo ; operculo depresse conico acuminulato ; peristomio parvo, dentibus pallidis subsubulatis, apice punctulatis, processibus apice punctulatis ciliisque in unum angustissimum conflatis in membrana usque ad dentium longitudinis ⅓ exserta impositis, annulo triplici circumdato.

Swain's Bay and Royal Sound, with fruit ripened, *Eaton*.

The very narrow leaves retain the same position in both the wet and dry state, they are narrower than observed in any form of *B. pendulum*.

Tab. III. f. iv. ; 1, natural size ; 2, cauline leaf; 3, leaf from perichætium ; 4, capsule; 5, portion of peristome ; all *magnified*.

7. **Bryum** (ECCREMOTHECIUM) **bimum,** *Schreb. ; Bryol. Europ.* t. 21. Christmas Harbour, *Hooker*. Near Swain's Bay ; and Royal Sound, with ripe fruit. *Eaton*.

The specimens vary in size, the stems in some being nearly three inches high, the lower leaves are all blackened.

8. **Bryum** (ECCREMOTHECIUM) **alpinum,** *Linn.* Royal Sound, with shining red foliage ; and Swain's Bay, all barren, *Eaton*.

The red-leaved specimens are exactly similar to those states of this species which are found in sub-alpine regions in Europe ; those states which are found in the plains have never the lustrous appearance which adorns this handsome moss.

The small specimen from Swain's Bay was mistaken for *B. lævigatum*, Hook. f. et Wils. (also a Kerguelen species), to which in colour it has a great resemblance, and the similarity was increased by the points of the upper leaves being broad and obtuse ; the lower leaves are, however, of the usual form.

9. **Bryum** (ECCREMOTHECIUM) **argenteum,** *Linn.* On sea cliffs near Observatory, barren, *Eaton*.

A small silvery state with the leaf points not produced.

10. **Bryum** (ECCREMOTHECIUM) **kerguelense,** *Mitt. in Journ. Linn. Soc.* xv. 67. Monoicum, cæspitosum. Caulis brevis, ramosus. Folia erecto-patentia, imbricata, inferiora rameaque ovali-lanceolata, acuta, carinato-concava, nervo rubro percursa, margine integerrimo, cellulis angustioribus in seriebus duabus limbum subindistinctum formantibus; reliquis suboblongis ; comalia longiora latioraque ; perichætialia interna minora. Theca in pedunculo breviusculo rubro superne flexuoso

curvato horizontalis, tenui-membranacea, nitida; sporangio ovali collo recto æqui-
longo sensim angustato; ore satis parvo coarctato; operculo convexo apice brevissimo acuto; peristomii dentibus pallidis interni fragmentis externo usque ad medium
adhærentibus.

Kerguelen Island, *Moseley*.

Stems including the numerous branches about 3 lines high, and with the
foliage about half a line wide. Leaves appressed when dry, a few at the apices of
the branches green, the lower all dark brown. Seta 3 lines long. Capsule about 1½
lines long, ochraceous, almost shining. The male flowers are terminal on branches
arising below the perichætium.

This small species appears to be nearly allied to *B. demissum*, Hook., but its
capsule is symmetrical, and the peristome is different.

Tab. III. fig. iii.; 1, plants nat. size; 2, entire plant; 3, cauline leaf; 4, perichætial and comal leaves; 5, portion of peristome; all *magnified*.

11. **Bryum lævigatum**, var. β., *Hook. f. and Wils. Flor. Antarel.*, 115,
t. 154, f. 3.

Christmas Harbour, barren, *Hooker*.

12. **Bryum Wahlenbergii**, *Schwæg*.

Christmas Harbour, common, *Hooker*.

[B. WARNEUM, Bland.; GAYANUM, Mont.; TORQUESCENS, Br. and Sch.; and
PALLESCENS, Schwæg., are all enumerated by James as found by Kidder (Bull. U. S.
Nat. Mus. 3, 26.]

1. **Mielichhoferia campylocarpa**, *Hook. el Arn. in Hook. Icon. Pl.*,
t. 136 (Weissia).

Kerguelen Island, *Moseley*. Near Swain's Bay, with unripe fruit, *Eaton*.

First described from the Andes, where it was gathered by Jameson; it was
afterwards found in Mexico, and may be one of those species extending throughout
the Andine chain. *M. basillaris*, Bruch et Schimp., from the Abyssinian mountains,
with entirely the same stature and appearance, differs in some particulars of the
peristomial teeth, and in the nerve of the leaf vanishing below the point.

Plagiothecium antarcticum, *Mitt. in Journ. Linn. Soc.* xv. 71. Monoicum, cæspitosum, ramis ascendentibus. Folia compressa, subfalcata, nitida; caulina
ovata, acuminata, integerrima, enervia; ramea ovato-lanceolata, tenuiter acuminata,
subenervia; omnia basi subcordata, cellulis angustis elongatis areolata; perichætialia
convoluta, late ovata, breviter acumiuata. Theca in pedunculo elongato rubro
ovalis, inæqualis, subcrecta inclinatave; operculo breviter conico; peristomio in-
terno ciliis in unum coalitis inter processus carinatos dentium longitudinis impositis
in membranam usque ad dentium dimidiam longitudinem exsertam insidentibus.

Royal Sound, with mature and old fruit, *Eaton*. Marion Island, *Moseley*.

Stems forming extensive soft patches, with shining foliage about half a line wide.
Seta about half an inch long, when dry twisted. Capsule obovate, the neck collapsing plicate, and so curved that the whole capsule is inclined; mouth large;

pale peristome prominent. The male flower, as is frequent in this genus, forms one of a cluster of small bud-like flowers at the base of the perichætium.

Closely resembles the European *P. nitidulum, Wahl.*, scarcely differing except in the base of its leaves. This is the species which is mentioned in Hooker's Handbook of the New Zealand Flora, ii. 476, as Hypnum pulchellum Dicks? from the Canterbury Alps.

Tab. III. Fig. v.; 1, plant nat. size; 2, cauline leaf; 3, perichætium with male flower at base; 4, capsule; 5, portion of peristome—all *magnified*.

[*P.* DONTANUM, *Sm.*, is enumerated by James as having been collected by Kidder in the U. S. Transit Expedition.]

1. **Acrocladium politum,** *Hook. f. et Wils. Fl. Antarct.* ii., t. 154, f. 2 (Hypnum). *Mitt.* l. c.

Hab.—Christmas Harbour, slender, tufted state, *Hooker*. Royal Sound, small and barren, *Eaton*.

This moss resembles some species of *Plagiothecium*, but seems to differ in habit, its branches with conduplicate bifarious leaves having so close a resemblance to those of *Phyllogonium elegans*, Hook. f. et Wils., that it is frequently mistaken for that plant. In the review of the genus *Orthorhynchum*, Reich. by C. Müller (Linnæa Band, 36, p. 28), one of the species to be referred to this genus, the *O. Hampeanum*, C. Muller, sent from Australia Felix by Baron F. von Mueller, must, from the description, be *Acrocladium politum*, of which specimens have been seen from Von Mueller, not however exactly corresponding in locality.

1. **Stereodon cupressiformis,** *Linn.* (Hypnum).

Base of sea cliff, Royal Sound, barren, *Eaton*.

The small specimens obtained exhibit this variable species in that form which in Europe is found on the roofs of buildings or on the ground; they are very unlike *S. chrysogaster*, C. Muller, so common in New Zealand.

1. **Amblystegium uncinatum,** *Hedw.*

Christmas Harbour, *Hooker*. Near Vulcan Cove, a tall robust form with nearly mature fruit; and Royal Sound, a similar state, but barren, *Eaton*.

2. **Amblystegium fluitans,** *Dill.*

West side of Swain's Bay, barren, *Eaton*.

A large state, with all but the terminal leaves of a brown colour.

3. **Amblystegium riparium,** *Linn.*

In the lake at Christmas Harbour, *Hooker*.

Specimen in a very imperfect state. Also found by the U. S. Transit Expedition growing with *Ranunculus crassipes*.

4. **Amblystegium kerguelense,** *Mitt.* Dioicum? Caulis decumbens, ramis confertis ascendentibus pinnatim ramosis. Folia caulina laxe imbricata, stricta vel curvata, ovato-lanceolata, subulato-acuminata, integerrima, nervo basi lato sensim angustato et in acumen evanido percursa; cellulis parvis oblongis limitibusque pellucidis ad angulos paucis rectangulis latioribus areolata; folia ramea

erecto-patentia, angustiora, nervo crassiore. Hypnum filicinum var. et H. serpens, var. *Flor. Antarct.*, p. 419 et 418.

Christmas Harbour, *Hooker*. Near Swain's Bay, barren, *Eaton*.

The single patch of this moss gathered by Mr. Eaton exhibits the species as very closely resembling *A. filicinum*, Linn., when it has not assumed a pinnate form; it is larger than *A. serpens*. The foliage is fulvous, neither wet nor dry is it altered in appearance.

5. **Amblystegium decussatum**, *Hook. f. et Wils. Fl. New Zeald.* ii. t. 90, f. 2. (Hypnum.)

Royal Sound, a slender straggling state, with irregular branches and an upright form, amongst *Bryum pendulum;* near Swain's Bay, an upright state more robust and more branched; near Vulcan Cove, a still larger state, with stems three inches high; all barren, *Eaton*.

All the specimens referred to this species have but little external resemblance to the complete state found fertile in New Zealand, but they agree very closely in the areolation of their leaves, and it is probable they are only slender forms similar to those produced by *A. filicinum*.

1. **Sciaromium conspissatum**, *Hook. f. et Wils. Fl. Antarct.* 419, t. 155, f. 3. (Hypnum).

Christmas Harbour; *Hooker, Moseley*. A short barren state.

All the Kerguelen specimens are smaller than those from the Falkland Islands.

1. **Brachythecium subpilosum**, *Hook. f. et Wils. Fl. Antarct.* 418, t. 154, f. 4. (Hypnum).

Kerguelen Island, *Moseley*.

More robust than the original specimens from Cape Horn, and in this respect nearer to the *Hypnum rutabulum*, var. 5, Fl. Antarct., from the Falkland Islands, which has since been named *H. subplicatum*, Hampe. If, however, the species may be supposed to vary as much in aspect as the European *B. rutabulum*, these slightly larger forms may be fairly considered within the limits of probable variation. Intermediate between the Hermite Island specimens and those from Kerguelen are some barren mosses from Otago, New Zealand, and some others collected in the Australian Alps by Von Mueller, to which it is probable the description of Dr. Hampe's *Hypnum austro-alpinum* may apply, as he says that the seta is thick and rough, and the capsule short, which are the most prominent characters appertaining to *B. subpilosum*.

2. **Brachythecium salebrosum**, *Hoffm.* (Hypnum). Hypnum rutabulum, var. 4, *Hook. f. et Wils. Fl. Antarct.* 417.

Christmas Harbour, *Hooker*. Hill N.W. of Mount Crozier, a fine silky state in large tufts, with stems 2–3 inches long; Swain's Bay, in boggy ground on the west side, a smaller state, all barren, *Eaton*.

This species is distinguished from *B. rutabulum* by the form of the leaves on the principal stems, which are not so dilated at their base, the outline being more nearly ovate and not deltoid. Specimens collected by Dr. Lyall in the Arctic regions at Beechy Island, correspond very nearly with the Kerguelen moss.

3. **Brachythecium paradoxum,** *Hook. f. et Wils. Fl. Antarct.* 449, t. 155, f. 2. (Hypnum).

Royal Sound, and Swain's Bay, with mature fruit, *Eaton*.

This species, which is found also in New Zealand and Fuegia, varies in size; the Kerguelen specimens are smaller than those from New Zealand; its affinity is with the European *B. relulinum* (Linn.), which is sometimes seen with falcate leaves, and then presents an appearance very different from its more usual state.

1. **Psilopilum trichodon,** *Hook. et Wils. in Hook. Lond. Journ. Bot.* vi. 289. (Polytrichum).

Hill N.W. of Mount Crozier, with narrow capsules, *Eaton*.

Originally described from the Andes of New Grenada, where it was found near the snow by Purdie; it was afterwards gathered by Jameson in a similar situation in the Andes of Quito.

Pogonatum alpinum, *Dill.*

Swain's Bay, with unripe fruit, *Eaton*.

This species occurs also in Australia, and has been described as *P. pseudoalpinum* (C. Müller, Bot. Zeit. 1855, 750), but it is admitted that the southern specimens differ scarcely if at all from those of the boreal regions.

[CATHARINA COMPRESSA, *C. Müll.*; Polytr. compressum, *Hook. f. et Wils.*, is enumerated amongst the United States Expedition collections.]

1. **Andræa acuminata,** *Mitt.* A. acutifolia, var. γ, *Hook. f. et Wils. Fl. Antarct.* 396.

Christmas Harbour, *Hooker.* Kerguelen Island, with a few mature capsules, *Moseley;* Royal Sound, without fruit, *Eaton*.

In the outline of its leaves this species resembles *A. marginata*, Hook. fil. et Wils. Fl. Antaret. 396, t. 151, f. 1., but the areolation of their upper portion is different, the cells being about $\frac{1}{3000}$ of an inch long by $\frac{1}{7000}$ wide, those in the corresponding portion of the leaves of *A. marginata* being about $\frac{1}{3000}$ wide, and $\frac{1}{500}$ long.

2. **Andræa squarrosa,** *Mitt. Musc. Austr. Amer.* 629. A. alpina var. 1, *Hook. f. et Wils. Flor. Antarct.* 395.

Christmas Harbour, *Hooker*.

This species has the perichœtial leaves in the Kerguelen specimens of the same form as in those collected by Prof. Jameson in the Andes of Quito.

[A. MARGINATA, *Hook. f. et Wils. Flor. Antarct.* 396, t. 151, fig. 1., has been found in Kerguelen Island by Kidder.]

III. *Hepaticæ.*

By WILLIAM MITTEN, A.L.S.

NINE species of *Jungermannia* and one *Marchantia*, were gathered by Dr. Hooker. These were arranged in 5 genera, and 5 of the species were described as new, the remainder being similar to species found elsewhere; none of the species were especially remarkable. Mr. Moseley collected at the time of the "Challenger's" stay 12 species, 7 of which were different from those obtained by Dr. Hooker, and 6 genera were also added to the flora. Fourteen species were found by Mr. Eaton; of these 8 species and 2 genera were additional to those previously known, bringing the whole number of the *Hepaticæ* up to 25.

The *Hepaticæ* of Kerguelen are allied most nearly to those of the Auckland and Campbell's Islands, and of Fuegia.

Noteroclada porphyrorhiza, *Leioscyphus pattens*, and *Teinnoma quadripartita*, are found also in Fuegia. *Jungermannia colorata*, and *Symphyogyna podophylla*, are found at the Cape of Good Hope. The former is, however, very widely distributed in austral regions. As with the mosses, it is remarkable how many additions were made to the flora by the small number of specimens obtained by each collector.

1. **Plagiochila heterodonta**, *Hook. f. et Tayl. Fl. Antaret.* 428, t. 157, f. 2.

Christmas Harbour, on moist rocks, *Hooker*. Royal Sound, barren, *Eaton*.

The specimens closely resemble those gathered by Dr. Hooker; it appears to be always a small species.

2. **Plagiochila minutula**, *Hook. f. et Tayl. Flor. Antaret.* 427, t. 157, f. 1.

Christmas Harbour, on rocks and the ground, *Hooker*.

1. **Leioscyphus turgescens**, *Hook. f. et Tayl. Fl. Antaret.* 150, t. 64, f. 2.

Hab. Royal Sound, amongst *Ditrichum Hookeri*, *Eaton*. (Lord Auckland's group).

2. **Leioscyphus pallens**, *Mitt. in Journ. Linn. Soc.* xv., 68. Caulis procumbens ascendensque, parce ramosus. Folia sursum secunda, conniventia, imbricata, orbiculata, caviuscula, integerrima, cellulis rotundis parietibus crassiusculis areolata. Amphigastria erecto-patentia, lanceolata, profunde bifida, laciniis elongatis subulatis. Folia involucralia majora, conformia; amphigastrio parvo quadrifido laciniis dentatis integerrimisve. Perianthium obovatum, ore truncato integerrimo.

Royal Sound, associated with *L. turgescens*, barren, *Eaton*.

Stems from an inch to an inch and a half long, seldom branched, with the leaves ½ line wide. Leaves pale olive-green, becoming in age brown, rather firm, not collapsing when dry, composed of rounded cells which at first contain small round granules that disappear in the older leaves. Stipules ½ line long, the one immediately under the perianth is small and easily overlooked. Perianth compressed. No capsuliferous stems have been seen.

It appears that in this species, and in some others of the same genus, the compressed truncate perianth is the result of the small size of the involucral stipule, which in the coalescence of the leaves of which the perianth is theoretically formed, is too small to affect its form, the reverse of which is so evident in the perianth of *Lophocolea*.

Tab. III., Fig. vi., plant *nat. size*; 2. leaf detached; 3. stipule from the stem; 4. perianth as seen laterally with involucral leaves; 5. stipule next the perianth; all *magnified*.

1. **Lophocolea pallidovirens**, *Hook. f. et Tayl. Fl. Antaret.* 430, t. 159, t. 9.

Kerguelen Island, *Moseley*. Near Vulcan Cove, barren, *Eaton*. (Fuegia).

2. **Lophocolea Novæ Zealandiæ**, *Lehm. et Lindenb.* (Jungermannia).

Royal Sound, fragments amongst *Ditrichum Hookeri*. Hill N.W. of Mount Crozier, with young perianths, *Eaton*. (New Zealand and Lord Auckland's group).

3. **Lophocolea humifusa**, *Hook. f. et. Tayl. Fl. Antaret.* 436, t. 159, f. v.

Christmas Harbour, *Hooker*; near Observatory Bay, barren, *Eaton*.

The specimens are pale yellowish green, and seem not different from *L. bidentata*, with which it agrees in perianth.

1. **Teinnoma quadripartita**, *Hook. Musc. Exot.* 117 (Jungermannia).

Kerguelen Island, a few small fragments, *Moseley*. Gathered also amongst *Dicrana* at Christmas Harbour by *Hooker*.

1. **Jungermannia cylindriformis**, *Mitt. in Journ. Linn. Soc.* xv. Exilis.

Caulis procumbens, ascendens, subsimplex, vix radicans. Folia subalterna, antice incurva, oblongo-ovalia, obtusa, sinu parvo obtuso obtuse bidentata, dentibus sæpe conniventer incurvis; involucralia minora, acute bidentata vel caulinis conformia. Perianthium elongatum, cylindraceum, obtusum, apice plicatum.

Royal Sound, in very small quantity with perianths amongst *Ditrichum Hookeri*, and hill N.W. of Mount Crozier, with *Seapania etandestina*, *Eaton*.

Stems about 2 lines long. Leaves ¼ line long, brownish green. Perianth 1 line long, of the same colour as the leaves. This minute plant is nearly related to *J. inflata*, Huds., having the same cylindrical perianth, and involucral leaves not much different from those of the stem, which are the characters of the genus *Gymnocolea*, Dumort, which comprises besides the European *J. inflata*, and the *J. turbinata*, Raddi.

Tab. III., Fig. vii.; 1, plant *nat. size*; 2 and 3, perianth and involucral leaves, dorsal and lateral view; 4, cauline leaf, expanded; all *magnified*.

** E

2. **Jungermannia leucorhiza,** *Mitt. in Journ. Linn. Soc.* xv. 68.   Caulis procumbens, radicellis pallidis.   Folia laxe inserta, quadrata subrotundave, sinu acuto obtusove bilobata, interdum lobo altero minore; lobis acutis obtusisve, incurvis; cellulis rotundatis et ovali-hexagonis areolata.

Kerguelen Island, in very small quantity amongst mosses, barren, *Moseley*. Stems less than 1 inch long, with the leaves ½ line wide.   Leaves green, tinged with brown.

Incomplete specimens of a species not before noticed in the Antarctic regions, but which appears to be near to the European *J. ventricosa*, Dicks, and to some states of *J. barbata*.

3. **Jungermannia colorata,** *Lehm. et Lindenb.*
Christmas Harbour, abundant on the hills, *Hooker, Moseley* (with perianths).

1. **Solenostoma humilis,** *Hook. f. et Tayl. Fl. Antarct.* ii. 434, t. 158, f. 6. (Jungermannia); J. inundata, *Flor. Nov. Zealand.* 128, t. 93, f. 3.

Hab. Christmas Harbour, barren, *Hooker*. A few fragments with one perianth, *Moseley*.

Both *S. humilis* and *J. inundata* were originally described as stipulate species, no amphigastria have, however, been since found on the specimens. It is probable that the figure of the supposed stipule of *J. humilis*, may have been drawn from a fragment of *Leioseyphus turgescens*.

Scapania, Lind (ex parte). Perianthium terminale, læve, a tergo ventreque compressum, ante capsulæ emissionem apice decurvum, herbaceo-membranaceum, ore truncato. Involucri folia 2, libera, caulinis conformia.—Plantæ terricolæ. Rami erecti ascendentesve, simplices vel furcati. Folia fere ubique æqualia, bifaria, equitantia, profunde biloba, laciniis subæqualibus apicibus rotundatis vel plus minus bifidis, textura e cellulis parvis. Amphigastria nulla.

This description is that of the Synopsis Hepaticarum, with slight modification, it applies to *S. densifolia, certebralis,* and *chloroleuca,* all so intimately related that the possibility of their being forms of one species may be conjectured. These differ from the chiefly European species which were included in the original idea of *Seapania,* and which are now by right of priority assigned to *Martinellia,* Gray, in having leaves not keeled in the space between the equal lobes, a peculiarity which gives the plants a different aspect. The perianth known from a single example on *S. vertebralis,* is like that found in *Martinellia,* but is narrowed upwards, truncate, the mouth bent over and denticulate.

1. **Scapania densifolia,** *Hook. Musc. Exot.* 36 (Jungermannia).
Kerguelen Island, *Moseley*.

The specimens agree with those gathered by Menzies, and are of the same brown colour. The distinction between *S. densifolia* and its congeners may be thus stated:—
S. densifolia, Hook., lobis foliorum apice integris rarius emarginatis.—*S. vertebralis,* Tayl., lobis apice exsectis.—*S. chloroleuca,* Hook. f. et Tayl., lobis apice bifidis.

2. **Scapania clandestina,** *Mont. Bol. Crypt. Astrolabe,* t. 16, f. 4. Balantiopsis incrassata, *Mitt. in Journ. Soc. Linn.* xv. 197.

Hill N.W. of Mt. Crozier, in very small quantity with *J. cylindriformis, Eaton.*

The stems of this small plant are about ½ inch high, and with the leaves ½ line wide. Leaves firm, with small round cells; lobes unequal and differing in their direction, the dorsal patent, the ventral nearly twice as large and divergent. In the Kerguelen specimens the space between the lobes is keeled and curved, and both the lobes are denticulate, except the superior edge of the ventral lobe which is only denticulate towards the apex, and like that of the dorsal lobe is terminated by two larger teeth (hence bidentate, with a small rounded sinus). In this particular they nearly resemble the leaves of *Balantiopsis diplophylla* and *B. erinacea,* Tayl. (Scapania), but differ in their dense areolation. No authentic specimen has been seen of *S. clandestina,* Mont., but the figure quoted agrees except in the areuation of the carina. A single stem picked from a tuft of *Aneura* from New Zealand has the lobes more nearly equal, the carina straight, very much longer, and all the marginal teeth more spiniform; it is probable as suspected in the Synopsis Hepaticarum, that the plant in a complete state would be different from the imperfect specimens yet seen. This species departs from *S. densifolia* and its allies in the leaves being carinate, and thus corresponds to *Martinellia;* it has, however, the apices of its leaves bedentate, which give it a different look from any of the species referred to that genus.

1. **Cesia atrocapilla,** *Hook. f. et Tayl. Fl. Antaret.* 423.

Foul haven, on clay banks, *Hooker;* in small blackish patches closely interwoven, *Moseley.*

From the examination of some branches of the specimens collected by Dr. Hooker it appears that fertile shoots would have their upper leaves nearly or quite entire and nearly orbicular in form.

1. **Lembidium ventrosum,** *Mitt. in Journ. Soc. Linn.* xv., 69. Caulis humilis, late compactoque cæspitosus, ascendens vel erectus, arcuatus, crassus, ramosus, innovationibus flagelliformibus ex amphigastriorum angulis emittens. Folia inferiora remota, superiora majora, insertione fere verticalia, patentia, apicem caulis versus imbricata, rotundata, profunde concava, apice rotundata sinuve subindistincto subrotusa, cellulis parvis parietibus angustis areolata. Amphigastria parva, cauli appressa, subtriangulari-ovata, apice submarginata. Perianthium in ramo superne valde incrassato, foliis amphigastrioque involucralibus convolutis ovatis apice breviter bitri-denticulatis, trigonum, ovatum, apice plicatum, ore laciniis conniventibus denticulatis obtusum.

Hill N.W. of Mt. Crozier, in dense tufts on the earth, with capsules just rising, *Eaton.*

In extensive brownish olive-green patches. Stems about 4 lines high, with the leaves scarcely ½ line wide, closely congested and cohering with very slender

E 2

hyaline rootlets. Perianth large for the size of the plant, arising from the apex of
a thickened branch; apex obtuse before the egress of the rather large spherical
capsule, but afterwards sub-truncate.   Spores minute, round, smooth, brown,
accompanied by fusiform moniliate fibres.

L. nutans, Hook. f. et Tayl. Fl. Nov. Zeald. 160, t. 65, f. 8, is a larger species,
and appeared by itself different from any genus that has been described, whereas *L.
centrosum* resembles the *Jungermannia Francisci* of Hooker's Brit. Jungermanniæ,
t. 49, a species which also produces thickly fleshy stolons, is irregular in the emar-
gination of its leaves, has the same kind of stipules as well as perianth, and
is therefore a species of *Lembidium*. How this genus or group of species may be
distinguished from the *Cephalozia* of Dumortier must remain for examination.

Tab. III., Fig. viii.; 1, plant of nat. size; 2, portion of stem with leaves and
stipules; 3, perianth and involucral leaves on lateral branch;—all *magnified.*

1. **Herpocladium fissum,** *Mitt. in Journ. Linn. Soc.* xv., 60.   Caulis per-
pusillus, firmus, crassiusculus.   Folia alterna, patentia, ovata, obtusa, apice incurva,
sinu parvo acuto breviter acute bidentata, concava, basi utroque caulem ad medium
diametrum tegentia, margine dorsali interdum flexura sinuata rarius unidentata,
cellulis densis obscuris areolata.   Amphigastria foliis similia, patula divaricatave,
apice obtusa, integra.

Kerguelen Island, *Moseley.*

Stems 3–4 lines long, with the leaves ¼ line wide.   The entire plant almost
black.

Tab. III., Fig. ix.; 1, plant of nat. size; 2, portion of stem, with leaves and
stipule from the dorsal side; 3, lateral view of leaf and spreading stipule; 4, leaf
detached and expanded;—all *magnified.*

1. **Tylimanthus viridis,** *Mitt. in Journ. Linn. Soc.* xv., 197.   Humilis.
Caulis erectus ascendensve, apice decurvus, subsimplex.   Folia distiche expansa,
oblongo-quadrata, apice oblique sinu lato subtruncato-bilobata; lobis obtusis, dorsali majore apicem versus interdum subdentatis, cellulis parvis rotundatis limitibus
pellucidis areolata.

Royal Sound, and hill N.W. of Mount Crozier, all barren, *Eaton.*

Stems ½ inch high, green, with the leaves scarcely 2 lines wide.   Leaves green,
frequently convex from the recurvation of the margin.   This nearly resembles *T.
tenellus,* Tayl. (*Gymnanthe*) from Tasmania, but it seems to be a smaller species.

Tab. III., Fig. x.; 1, plant of *nat. size;* 2, portion of stem with leaves
enlarged.

1. **Marsupidium excisum,** *Mitt. in Journ. Linn. Soc.* XV., 69.   Caulis pri-
marius repens, exinde ascendens, pallidus.   Folia inferiora minora, deinde superiore
minora, omnia oblongo-quadrata, concava, sinu obtuso bidentata, integerrima sub-
crenatave, lobis latis acutis incurvis, cellulis protuberantibus papulosa.

Royal Sound, with *Acrocladium politum* and *Pogonatum alpinum,* barren, *Eaton.*

Primary stems of the same colour as the leaves, fleshy, obscure, creeping; from these arise erect or ascending simple or branched shoots, which are arcuate, their points attenuated and decurved. The leaves where largest are about ½ line long, and when flattened of the same width, of a pale obscure olive-green; bases not decurrent; insertion variable but generally oblique; margins entire, or obtusely sub-crenate; areolation of hexagonal or rounded cells with thin walls, enclosing a few green granules and projecting on both surfaces, but most on the external, as hyaline papulæ. Papulæ of the same kind are also present on the younger stems, but less prominent.

No kind of inflorescence has been seen on this species, and its location here is conjectured from its having the same habit as *M. Knightii*, from New Zealand.

Tab. III., Fig. xi.; 1, plant of *nat. size;* 2, part of stem with leaves; 3, cells from middle of leaf; both *magnified.*

1. **Fossombronia australis,** *Mitt. in Journ. Soc. Linn.* XV., 73. Caulis cæspitosus, prostratus vel ascendens, arcuatus, radicellis purpureis. Folia sub-quadrata, angulata, margine flexuosa, antice incurva. Perianthium turbinatum, margine flexuosum, angulatum, semina rotunda limbo hyalino lævia.

Kerguelen Island, *Moseley.* Royal Sound, and near Vulcan Cove, with young capsules, *Eaton.* (Heard Island, *Moseley*).

Some of the specimens are very large, with arcuate stems more than an inch long, producing many purple rootlets. The leaves are 1 line long by about 1½ wide, green, with pellucid cells.

2. **Fossombronia pusilla,** *Linn.*

Christmas Harbour, amongst moss, *Hooker.*

1. **Noteroclada porphyrorhiza,** Nees; N. confluens, *Fl. Antarct.* 446, t. 161, f. 7.

Christmas Harbour, on moist banks, *Hooker.*

1. **Symphogyna podophylla,** *Thunb.* (Jungermannia); *Gottsche, Lindenb. et Nees Syn. Hepat.* 481.

Near Vulcan Cove; Royal Sound; hill N.W. of Mount Crozier; all without fructification, *Eaton.*

1. **Aneura multifida,** *Linn.* (Jungermannia).

West side of Swain's Bay, on boggy ground, and near Vulcan Cove, all barren, *Eaton.*

2. **Aneura pinguis,** *Linn.* (Jungermannia).

West side of Swain's Bay, small and barren, *Eaton.*

1. **Marchantia polymorpha,** Linn.

Christmas Harbour; *Hooker, Moseley;* Royal Sound and Swain's Bay, *Eaton.* All the specimens produce scyphi, but are otherwise barren.

V.—*Lichens.*

By the Rev. J. M. Crombie, F.L.S.

The first record that we can find of the Lichen-flora of this remote island, is contained in a preliminary account of the Antarctic Lichens collected by Dr. J. D. Hooker [*] during the voyage of the "Erebus" and "Terror," which was published by him and Dr. Thomas Taylor in the "London Journal of Botany" (1844), vol. iii. pp. 634–658. The Kerguelen Island lichens there enumerated amount in number to 17 species, named by Dr. Taylor; but at least one half of the names attributed to them are misapplied, and therefore must be excluded, owing chiefly to the determinations having been attempted in the absence of such microscopical analysis of the specimens as is now found to be essential for their discrimination. The number was subsequently raised to 27 species and varieties, when the list was revised by the Rev. Churchill Babington for publication in Dr. Hooker's "Flora Antarctica" (1847), vol. ii. pp. 519–542. A considerable proportion of the names in this later list must however be rejected for the same reason as those erased from the previous one. Unfortunately authentic examples of several of Dr. Taylor's critical species are wanting in the Kew Herbarium; [†] and his collection (now in the Herbarium of the Boston Society of Natural History), according to Professor Edw. Tuckerman, contains very little that is illustrative of his Kerguelen Island determinations. I have lately published a further revision of the Kerguelen Island Lichens collected by Dr. Hooker, in the "Journal of Botany" for April 1877, wherein the number of the species is reckoned to be 18 or 19 besides 2 named forms.

Mr. Moseley of the Challenger Expedition gathered in this island upwards of 13 species and 1 named form. (*Vide* Crombie in Journ. Linn. Soc. Bot. 1877.)

Dr. Kidder of the American Transit of Venus Expedition collected in the vicinage of Molloy Point 13 or 14 species and 1 named form. These with others from the Taylor collection are specified by Prof. Ed. Tuckerman in Bulletin U.S. Nat. Mus. No. 3 (1876), and are noticed by me in the "Journal of Botany" for April 1877.

The collection made by Mr. Eaton between the end of October 1874 and the end of February 1875, in the district immediately to the westward of Dr. Kidder's station, comprises 50 or 51 species and 9 named forms. Of these about 30 were

---

[*] One (or more) species of Lichens was obtained in Kerguelen Island in 1776 by Mr. Anderson, the Surgeon and Naturalist who accompanied Captain Cook.—A. E. E.

[†] Dr. Taylor died shortly after the publication of his first rough determination of the Antarctic Lichens, and it was impossible to recover from the heap of his unarranged materials, which were in a confused state, all of the specimens which should have been returned. I strongly suspect, from the state of his notes sent to me from time to time, that he did not attend sufficiently to localities, and that some of the specimens in the Herbarium labelled as from Kerguelen Island did not come from that island.—J. D. H.

described as new species referred to known genera, in the Journal of Botany for November 1875 and January 1876, and again with fuller diagnoses with which Dr. Nylander (who most kindly assisted me in their determination) favoured me in the Journal of the Linnæan Society (Botany) for July 1876. Though several of the new species bear a superficial resemblance to some of our northern lichens, yet on analysis they are found to be quite distinct, and for the most part are peculiar to Kerguelen Island.

The results obtained by the German Transit of Venus and Surveying Expedition at Betsy Cove, are not yet published.

The total number of species obtained from the island is 61, and 10 varieties. Traces of a few other species exist in the various collections, consisting either of sterile thalli or undeveloped apothecia which are necessarily indeterminable.

1. **Lichina antarctica,** *Cromb. in Journ. Bot.* v. 21 (1876) ; *et in Journ. Linn. Soc.* xv. 181.

Observatory Bay, on dry rocks near the sea, *Eaton.*

1. **Amphidium molybdophlacum,** *Nyl. in Journ. Bot.* iv. 333 (1875) (errone molybdophœun); *Cromb. in Journ. Linn. Soc.* xv. 181. *et in Journ. Bot.* vi. 103, 106. Lecanora melanaspis, *Bab. in Flor. Antarct.* 536 (*exel. Syn.* L. dichron) Pannaria glaucella, *Tuckerm. in Bull. U. S. Nat. Mus.* iii. 28.

On earth and stones, Christmas Harbour, *Hooker ;* on stones in wet places, Swain's Harbour, *Eaton ;* Molloy Point, *Kidder.*

1. **Stereocaulon cymosum,** *Cromb. in Journ. Linn. Soc.* xv. 182, *et in Journ. Bot.* vi. 103. S. corallinum, *Hook. f. et Tayl. Flor. Antarct.* 528.

On rocks, altitude 6–1200 feet, Christmas Harbour, *Hooker, Moseley ;* top of a hill on west side of Carpenter's Cove, barren, *Smith Dorien (Eaton).*

1. **Cladonia fimbriata,** *Hoffm. ; Cromb. in Journ. Linn. Soc.* xv. 182. C. pyxidata, *Linn. ; Tuckerm. in Bull. Torr. Bot. Club, October* 1875, *et Bull. U. S. Nat. Mus.* 3, 29.

Dry slopes, Swain's and Observatory Bays, *Eaton ;* Molloy Point, *Kidder.*

VAR. COSTATA, Flk. Observatory Bay, *Molloy.*

2. **Cladonia cornuta,** *Linn. ; Cromb. in Journ. Linn. Soc.* 1877. Kerguelen Island, *Moseley.*

3. **Cladonia acuminata,** *Ach. ; Cromb. in Journ. Linn. Soc.* xv. 182. C. phyllophora, *Tayl.*

Christmas Harbour, *Hooker ;* observatory Bay, common, but sparingly fertile, *Eaton.*

1. **Neuopogon melaxanthus,** *Ach.* (Usnea); *Cromb. in Journ. Linn. Soc.* xv. 182; *et in Journ. Bot.* vi. 103, 106 (1877). Usnea sulphurea, *Müll. ; Tuckerm. in Bull. Torr. Bol. Club,* 1875, *et in Bull. U. S. Nat. Mus.* 3, 27. Ramalina scopulorum ε, *Hook. f. et Tayl. Flor. Antarct.* 522.

Rocks and boulders, Christmas Harbour, *Anderson, Hooker;* on the upper slopes and tops of hills, *Eaton.*

VAR. SOREDIIFER, *Cromb. l. c.*   Common at all altitudes, *Eaton.*

VAR. CILIATA, *Nyl.   Crombie, l. c.*   Observatory Bay, *Eaton.*

2. **N. Taylori,** *Hook. f.; Flor. Antarct.* 521, t. 195, *Fig.* i. (Usnea); *Cromb. in Journ. Linn. Soc.* xv. 183, *et in Journ. Bot.* vi. 103 (1877).

Rocks ascending to 1,200 feet; Christmas Harbour, *Hooker, Moseley;* Swain's Bay and Carpenter's Cove (but not near Observatory Bay), *Eaton;* Molloy Point, *Kidder.*

1. **Parmelia stygioides,** *Nyl.; Cromb. in Journ. Bot.* iv. 333 (1875), *et in Journ. Linn. Soc.* xv. 183.

Dry rocks and stony slopes, Swain's Bay, *Eaton.*

1. **Peltigera rufescens,** *var.* spuria, *DC.; Cromb. in·Journ. Linn. Soc.* xv. 183; *Journ. Bot.* vi. 103.   Peltidea venosa, *Hook. f. et Tayl. Flor. Antarct.* 525.

On wet moss, &c., Christmas Harbour, *Hooker;* Swain's Bay, *Eaton.*

2. **Peltigera polydactyla,** forma HYMENINA, *Ach.; Cromb. in Journ. Linn. Soc.* xv. 183.   P. polydactyla, *Flor. Antarct.* 524.   P. horizontalis, *Ach.; Flor. Antarct. l.c.* 525.

Amongst moss, &c., Christmas Harbour, *Hooker;* Observatory Bay, *Eaton.*

1. **Pannaria dichroa,** *Hook. f. et Tayl. in Lond. Journ. Bot.* iii. 643; *Cromb. in Journ. Bot.* vi. 106 (1877), *et in Journ. Linn. Soc.* xvi.   P. Taylori, *Tuckerm. in Bull. Torr. Bot. Club, October* 1875, *et in Bull. U. S. Nat. Mus.* 3, 28.   P. placodicopsis, *Nyl. in Journ. Bot.* iv. 334 (1875); *Cromb. in Journ. Linn. Soc.* xv. 183.   Lecanora melanaspis, *Ach.; Hook. f. et Tayl. Flor. Antarct.* 536.

On rocks, Christmas Harbour, *Hooker, Moseley;* Observatory Bay, sparingly, *Eaton;* Molloy Point, *Kidder.*

2. **Pannaria obscurior,** *Nyl.; Cromb. in Journ. Bot.* iv. 334 (1875), *et in Journ. Linn. Soc.* xv. 183.

On decayed moss, Observatory Bay, *Eaton.*

1. **Psoroma hirsutulum,** *Nyl.; Cromb. in Journ. Bot.* iv. 333 (1875), *et in Journ. Linn. Soc.* xv. 184.

On moss and dead stems of *Acæna,* very local, Observatory Bay, *Eaton.*

1. **Lecanora** (PLACOPSIS) **gelida,** *Linn.; Flor. Antarct.* ii. 535; *Tuckerm. in Bull. Torr. Bot. Club, Oct.* 1875, *et in Bull. U. S. Nal. Mus.* 3, 29; *Cromb. in Journ. Linn. Soc.* xv. 184, *et in Journ. Bot.* vi. 104, 106 (1877).

On stones, Christmas Harbour, *Hooker;* Observatory Bay, *Eaton;* Molloy Point, *Kidder.*

Var. VITELLINA, *Bab. in Flor. Antarct.* l. c.; *Cromb.* l. c.   Christmas Harbour, *Hooker, Moseley.*

Var. LATERITIA, *Nyl.; Cromb.* l. c.   Placodium bicolor, *Tuckerm.* l. c.   Christmas Harbour, *Hooker, Moseley.*   Swain's Bay and Royal Sound, *Eaton.*

LICHENS.—J. M. CROMBIE. 41

2. **Lecanora** (Placopsis) **macropthalma,** *Hook. f. et Tayl. in Lond. Journ. Bot.* iii. 660 (Urceolaria); *Tuckerm. in Butl. Torr. Bot. Club, Oct.* 1875, *et in Bull. U. S. Nat. Mus.* 3, 29; *Cromb. in Journ. Linn. Soc.* xv. 185 and xvi., *et in Journ. Bot.* vi. 104.

On stones in moist places, *Hooker, Eaton, &c.*

Lecanora (Placodium) **elegans,** *Ach.; Cromb. in Journ. Linn. Soc.* xv. 184, *et in Journ. Bot.* vi. 104 (1877); *Tuckerm. in Bull. U. S. Nat. Mus.* 3, 28. L. murorum, var. β, *Flor. Antarct.* 535.

On rocks and stones, Christmas Harbour, *Hooker;* Observatory and Swain's Bays, *Eaton.*

Var. lucens, *Nyl.; Cromb.* l. c. On dead stems of *Acœna* and *Pringlea,* Observatory Bay, *Eaton.*

4. **Lecanora subunicolor,** *Nyl.; Cromb. in Journ. Bot.* v. 19 (1876), *et in Journ. Linn. Soc.* xv. 184.

On rocks, Royal Sound, very sparingly, *Eaton.*

5. **Lecanora vitellinella,** *Nyl.; Cromb. in Journ. Bot.* iv. 334 (1875), *et* vi. 104 (1877), *et in Journ. Linn. Soc.* xv. 184. L. candelaria, *Ach.; Flor Antarct.* 537.

Maritime rocks, Christmas Harbour, *Hooker;* Observatory and Swain's Bays, *Eaton.*

6. **Lecanora cyphelliformis,** *Cromb. in Journ. Linn. Soc.* xvi.

Christmas Harbour, *Moseley.*

7. **Lecanora diphyella,** *Nyl.; Cromb. in Journ. Bot.* v. 21, *et in Journ. Linn. Soc.* xv. 184.

On rocks at low elevations, Observatory Bay, *Eaton.*

8. **Lecanora atro-cæsia,** *Nyl.; Cromb. in Journ. Bot.* iv. 334 (1875), *et* vi. 104 (1877), *et in Journ. Linn. Soc.* xv. 185. L. confluens, *Hook. f. et Tayl. in Lond. Journ. Bot.* iii. 636. L. albo-cœrulescans, *Ach.? Bab. in Flor. Antarct.* 538.

Rocks at Christmas Harbour, *Hooker;* Observatory Bay and stony slopes at Volage Bay, plentiful, *Eaton.*

9. **Lecanora brocella,** *Nyl.; Cromb. in Journ. Bot.* v. 21 (1876), *et* vi. 104 (1877), *et in Journ. Linn. Soc.* xv. 185.

On dead moss, &c., Christmas Harbour, *Hooker;* Observatory Bay, *Eaton.*

10. **Lecanora umbrina,** *Ach.; Cromb. in Journ. Linn. Soc.* xv. 185.

On dead plants, Observatory Bay, *Eaton.*

11. **Lecanora kergueliensis,** *Nyl.; Cromb. in Journ. Bot.* vi. 106 (1877). Urceolina kergueliensis, *Tuckerm. in Bull. U. S. Nat. Mus.* 3, 29.

Rocks at Molloy Point, *Kidder.*

12. **Lecanora sublutescens,** *Nyl.; Cromb. in Journ. Bot.* v. 21 (1876), *et in Journ. Linn. Soc.* xv. 186.

** F

On a shaded sea cliff near Observatory Bay, colouring the rock, *Eaton*.

[L. CITRINA, *Ach.*, L. ERYTHROCARPA, *Fr.*, and L. HAGENI, *Ach.*, enumerated in the Flora Antarctica, 536, from very imperfect materials, are too doubtful to be enumerated.]

1. **Pertusaria perrimosa,** *Nyl.; Cromb. in Journ. Linn. Soc.* xv. 186; *Journ. Bot.* vi. 104. P. communis, *DC.; Flor. Antarct.* 540. ? Lecanora tartarea, *Ach.; Flor. Antaret.* 536.

On rocks, Christmas Harbour, *Hooker;* Observatory and Swain's Bays, *Eaton.*

2. **Pertusaria subferruginosa,** *Cromb. in Journ. Linn. Soc.* xv. 186.

On rocks, Observatory Bay, *Eaton.*

3. **Pertusaria cineraria,** *Nyl.; Cromb. in Journ. Linn. Soc.* xv. 186.

On rocks, Volage and Swain's Bays, *Eaton.*

1. **Lecidea variatula,** *Nyl.; Cromb. in Journ. Linn. Soc.* xv. 186.

On dead stems of *Aeæna*, Observatory Bay, *Eaton.*

2. **Lecidea inundata,** *Fr.; Cromb. in Journ. Bot.* vi. 106 (1877). Biatora rubella, *Ehrh.; Tuckerm. in Bull. U. S. Nat. Mus.* 3, 29.

Molloy Point, *Kidder.*

3. **Lecidea assimilata,** *Nyl.; Cromb. in Journ. Linn. Soc.* xv. 187, *et in Journ. Bot.* vi. 104. L. aromatica *Ach. in parte Flor. Antaret.* 538.

Christmas Harbour, *Hooker;* Observatory Bay in turf, *Eaton.*

4. **Lecidea aromatica,** *Ach.; Nyl. in Journ. Bot.* vi. 104 (1877). *Flor. Antaret.* 538.

Christmas Harbour, *Hooker.*

5. **Lecidea enteroleuca,** *Fries ?; Tuckerm. in Bull. Torr. Bot. Club,* Oct. 1875, *et in Bull. U. S. Nat. Mus.* 3, 30. *Cromb. in Bot. Journ.* vi. 106 (1877).

On dead grasses, Molloy Points, *Kidder.*

6. **Lecidea assentiens,** *Nyl.; Cromb. in Journ. Bot.* iv. 334, *et* vi. 105 (1877), *et in Journ. Linn. Soc.* xv. 187. L. contigua *var.* hydrophila, *Bab. in Flor. Antaret.* 538.

On rocks, Christmas Harbour, *Hooker;* Observatory Bay, *Eaton.*

7. **Lecidea intersita,** *Nyl.; Cromb. in Journ. Linn. Soc.* xv. 187.

On rocks, Observatory Bay, very sparingly ; also ? one mile N.W. of Mount Crozier, *Eaton.*

8. **Lecidea phæostoma,** *Nyl.; Cromb. in Journ. Bot.* iv. 334 (1875), *et in Journ. Linn. Soc.* xv. 187.

On stones and bare soil, Observatory Bay, sparingly, *Eaton.*

9. **Lecidea amylacea,** *Ach.; Cromb. in Journ. Linn. Soc.* xv. 188 (1876), *et Journ. Bot.* vi. 104. L. spilota *Ach.; Bab. in Flor. Antaret.* 538. L. rivulosa, *Tayl. in Linn. Journ. Bot.* iii. 636.

On rocks and stones, Christmas Harbour, *Hooker;* Volage and Swain's Bays, *Eaton.*

10. **Lecidea subassentiens,** *Nyl.; Cromb. in Journ. Bot.* v. 21 (1876), *et in Journ. Linn. Soc.* xv. 188.

On rocks, Observatory Bay, very sparingly, *Eaton.*

11. **Lecidea perusta,** *Nyl.; Cromb. in Journ. Bot.* iv. 334 (1875), *et* vi. 105, 106 (1877), *et in Journ. Linn. Soc.* xv. 188. L. fusco-atra, *Ach.; Ftor. Antarct.* 539; *Tuckerm. in Butt. Torr. Bot. Club, et in Butt. U. S. Nat. Mus.,* 3, 30.

On rocks, Christmas Harbour, *Hooker ;* Observatory Bay, *Eaton ;* Molloy Point, *Kidder.*

12. **Lecidea asbolodes,** *Nyl.; Cromb. in Journ. Bot.* v. 21 (1876); *et in Journ. Linn. Soc.* xv. 188.

On rocks, Observatory Bay, *Eaton.*

13. **Lecidea lygomma,** *Nyl.; Cromb. in Journ. Bot.* iv. 334 (1875), *et in Journ. Linn. Soc.* xv. 189.

On rocks, Observatory Bay, *Eaton.*

14. **Lecidea subcontinua,** *Nyl.; Cromb. in Journ. Linn. Soc.* xv. 189, *et in Journ. Bot.* vi. 104, 106 (1877). Urccolaria endochlora, *in parte Ftor. Antarct.* 537.

On rocks and stones, Christmas Harbour, *Hooker ;* Swain's Bay, *Eaton.*

Var. FERREA, *Nyl.; Cromb. t. c.* Swain's Bay, *Eaton.*

15. **Lecidea Eatoni,** *Cromb. in Journ. Bot.* iv. 334, 1875, *et in Journ. Linn. Soc.* xv. 189.

On rocks and boulders, Observatory, Volage, and Swain's Bays, *Eaton.*

16. **Lecidea homalotera,** *Nyl. mss.; Cromb in Journ. Bot.* vi. 105 (1877). Urccolaria endochlora, *Hook. f. et Tayt. in part. Ftor. Antarct.* 537.

On rocks, Christmas Harbour, *Hooker.*

17. **Lecidea disjungenda,** *Cromb. in Journ. Bot.* vi. 105. Urccolaria endochlora, *Hook. f. et Tayt. in part. Ftor. Antarct.* 537.

On rocks, Christmas Harbour, *Hooker.*

18. **Lecidea subplana,** *Nyl.; Cromb. in Journ. Bot.* iv. 334 (1875), *et in Journ. Linn. Soc.* xv. 189.

On boulders, sparingly, Observatory and Swain's Bays, *Eaton.*

19. **Lecidea stephanodes,** *Strn.; Cromb. in Journ. Linn. Soc.* 1877. Kerguelen Island, *Mosetcy.*

20. **Lecidea Dicksonii,** *Ach.; Cromb. in Journ. Linn. Soc.* xv. 190. L. sincerula, *Nyl.; Cromb. in Journ. Bot.* v. 22 (1876).

On rocks, Royal Sound, Observatory, Volage, and Swain's Bays, *Eaton.*

21. **Lecidea tristiuscula,** *Nyl.; Cromb. in Journ. Linn. Soc.* xv. 190.

On stems, Volage Bay, sparingly, *Eaton.*

22. **Lecidea superjecta,** *Nyl.; Cromb. in Journ. Linn. Soc.* 1877. Kerguelen Island, *Mosetcy.*

23. **Lecidea myriocarpa,** *DC.; Cromb. in Journ. Linn. Soc.* xv. 190, *et in Journ. Bot.* vi. 106 (1877). ? Buellia parasema, *Tuckerm. in Bull. Torr. Bot. Club, et in Bull. U. S. Nat. Mus.* 3, 30.

On rocks, Swain's Bay, sparingly, *Eaton.* Molloy Point, *Kidder.*

Var. ERUMPENS, *Crombie, l. e.* On dead *Acæna* stems, Observatory Bay (a single specimen), *Eaton.*

24. **Lecidea subplicata**, *Nyl.; Cromb. in Journ. Bot.* iv. 334 (1875), *et in Journ. Linn. Soc.* xv. 190.

On rocks, Observatory and Swain's Bays, *Eaton.*

25. **Lecidea cerebrinella**, *Nyl.; Cromb. in Journ. Bot.* v. 22 (1876), *et in Journ. Linn. Soc.* xv. 191.

On rocks, Observatory Bay, *Eaton.*

26. **Lecidea stellulata**, *Tayl. in Flor. Hibern.* 118; *Flor. Antaret.* 539; *Tuckerm. in Bull. Torr. Bot. Club, et in Bull. U. S. Nat. Mus.* 30; *Cromb. in Journ. Bot.* vi. 105 (1877).

On rocks, Christmas Harbour, *Hooker;* near Molloy Point, *Kidder.*

27. **Lecidea geographica**, *Linn.; Cromb. in Journ. Linn. Soc.* xv. 191, *et in Journ. Bot.* vi. 105 (1877). L. geographica, *var.* urceolata, *Schærer; Bab. in Flor. Antaret.* 539. Buellia geographica, *Tuckerm. in Bull. Torr. Bot. Club, et in Bull. U. S. Nat. Mus.* 3, 30.

On rocks frequent, *Hooker, Moseley, Eaton, Kidder.*

1. **Verrucaria tessellatula**, *Nyl.; Cromb. in Journ. Bot.* iv. 335 (1875), *et in Journ. Linn. Soc.* xv. 191.

On rocks and stones, Volage and Swain's Bays, and (where overflown by the tides) at Observatory Bay, *Eaton.*

2. **Verrucaria obfuscata**, *Nyl.; Cromb. in Journ. Bot.* v. 22 (1876), *et in Journ. Linn. Soc.* xv. 191.

On stones, hill N.W. of Mount Crozier, *Eaton.*

3. **Verrucaria æthiobola**, *Ach.; Cromb. in Journ. Linn. Soc.* xv. 193.

On rocks, Observatory Bay, *Eaton.*

4. **Verrucaria chlorotica**, *Ach.; Tuckerm. in Bull. Torr. Bot. Club, et in Bull. U. S. Nat. Mus.* 3, 30. (Sagedia); *Cromb. in Journ. Bot.* vi. 106 (1877). Molloy Point, *Kidder.*

5. **Verrucaria prævalesens**, *Nyl.; Cromb. in Journ. Bot.* xv. 192. Rocks at Observatory Bay, and ? hill N.W. of Mount Crozier (*Eaton*).

6. **Verrucaria kerguelina**, *Nyl.; Cromb. in Journ. Bot.* v. 22 (1876), *et in Journ. Linn. Soc.* xv. 192.

On rocks, Observatory Bay, sparingly, *Eaton.*

7. **Verrucaria insueta**, *Nyl.; Cromb. in Journ. Linn. Soc.* xv. 192.

On rocks and stones, Volage Bay, *Eaton.*

8. **Verrucaria congestula**, *Strn.; Cromb. in Journ. Linn. Soc.* 1877. Kerguelen Island, *Moseley.*

[ISIDIUM OCULATUM, I. LUTESCENS, and LEPRARIA FLAVA, all enumerated in the Flora Antarctica as doubtful, are imperfect states of Lichens.]

V.—*Marine Algæ* (*exclusive of the Diatomaceæ*).

By G. Dickie, A.M., M.D., F.L.S., Professor of Botany in the University of Aberdeen.

The total number of marine species of *Algæ* known to be indigenous to Kerguelen Island (excluding *Diatomaceæ*) is 71. The collections upon which this estimate is based are those made respectively by—

Dr. Hooker (Antarctic Expedition) in the winter of 1840 (May—July), chiefly at Christmas Harbour, comprising 39 species;

Mr. Moseley (Challenger Expedition) in the summer of 1874 (January and February), chiefly at Christmas Harbour and the eastern coast as far as Betsy Cove, comprising 37 species;

Dr. Kidder (American Transit of Venus Expedition) in the spring and first part of the summer of 1874-5 (Sept. to Jan.) near Molloy Point, towards the entrance of Royal Sound, comprising 22 species;

And the Rev. A. E. Eaton (English Transit of Venus Expedition) in the spring and summer of 1874-5 (Oct. 11—Feb. 27), in the interior of Royal Sound (Observatory Bay) and in Swain's Bay, comprising 53 species.

The botanical results of the German Transit of Venus and Surveying Expedition, which was stationed for about two months at Betsy Cove, are not yet made known.

Mr. Eaton was at Observatory Bay during October, November, most of December, and the whole of February, during which time he made frequent use of the grapple. In Swain's Bay he collected *Algæ* on nine occasions between the 15th and the 30th of January inclusive. Of the 53 species in his collection 44 were obtained in Swain's Bay, and only 32 at Observatory Bay : 24 species (probably 27 or 28, *vide infra*) are common to both of the areas, 21 occurred (to Mr. Eaton) only in Swain's Bay, and 8 (from which 3 or 4 should be deducted, and added to the species common to both) were collected only in Observatory Bay. The preponderance of the Swain's Bay gatherings may partly be accounted for by the distance of Observatory Bay from the open sea. For Mr. Eaton noticed that in some very retired parts of Swain's Bay the components of the Alga flora and their state of growth were very similar to those prevailing at Observatory Bay. In advancing from the more sheltered to more open waters he observed considerable regularity maintained in the rate of change proceeding in the composition of the Alga flora ; so that it was possible, while collecting in one place, to conjecture beforehand with tolerable accuracy the number of additional species that would be found in other positions more exposed to the slight swell that enters the bay from the outer sea. And he was of opinion that if it had been possible to have visited the coast external to the bay, 10 or 12 species would most likely have been added to the 53 in his collection. Judging from the number of species apparently indigenous to un-sheltered situations which go to form the 18 that are not represented in his collec-tion, this conjecture may have been not far from the mark.

But the advantages of situation afforded in Swain's Bay for the growth of various Algæ absent from the almost waveless shores of Observation Bay would have availed nothing, had it not been for the liberality and kindness of Captain Fairfax, R.N., in command of H.M.S. "Volage." Having invited Mr. Eaton to be his guest for three weeks, he conveyed him in his gig to almost every part of the bay that was accessible by boat in Kerguelen Island weather, and surrendered his cabins without reservation to the reception of buckets and specimens of all descriptions, excluding only seals and cetacea accommodated elsewhere.

The local distribution of the species round the coast may be ascertained roughly from a comparison of the constituents of the collections above mentioned. Of the 71 species, 14 are common to all of the collections, and 8 common to three out of the four, making together 22 species, which may be regarded as plants generally distributed round the island; 14 are common to Mr. Eaton's collection and one of the other three, and 1 species to Dr. Hooker's and Dr. Kidder's,—together making 15 local plants, mostly of frequent occurrence; 5 are common to Dr. Hooker's collection and Mr. Moseley's (gathered in Christmas Harbour), and 29 are in one of the collections only, making 34 scarce or rare species. Of the 29, there are in Dr. Hooker's collection 7 species, in Mr. Moseley's 4, in Dr. Kidder's 1, and in Mr. Eaton's 17.

As to their general geographical range, 20, or rather more than a quarter of them, are found in various parts of the shores of Europe, and some are cosmopoli-tan. The following 8, so far as is known, are peculiar to the island :—*Desmarestia chordalis, Sphacelaria corymbosa* and *S. affinis, Melobesia kerguelena, Nitophyllum fusco-rubrum, Epymenia variolosa, Ptilota Eatoni,* and *Callithamnion simile.*

The following are the numbers of the species after their respective families :—

Fucaceæ, 2.	Sphærococcoideæ, 8.
Sporochnaceæ, 4.	Gelidiaceæ, 1.
Laminariaceæ, 2.	Rhodymeniaceæ, 4.
Dictyotaceæ, 2.	Cryptonemiaceæ, 11.
Chordariaceæ, 3.	Ceramiaceæ, 7.
Ectocarpaceæ, 3.	Siphonaceæ, 3.
Rhodomelaceæ, 4.	Ulvaceæ, 5.
Laurenciaceæ, 2.	Confervaceæ, 7.
Corallinaceæ, 3.	

Of these 16 belong to the Olive, 40 to the Red, and 15 to the Green Series.

There are also included in the present paper, for convenience, 4 freshwater species:—*Bostrychia caga, Vaucheria Dillwynii, Ulca cristata*, and *Prasiola fluviatilis.*

1. **D'Urvillea utilis,** *Bory; Flor. Antarct.* 454; *Dickie in Journ. Linn. Soc.* xv. 43, 198; *Farlow in Bull. U. S. Nat. Mus.* 3, 30.

On exposed rocks at and below half-tide level, not in very sheltered situations; abundant.—In the Southern ocean, between lat. 45° and 55° S., reaching to lat. 65° S. in the meridian of New Zealand (Hooker).

2. **D'Urvillea Harveyi,** *Hook. f. Flor. Antarct.* 456, t. clxv., clxvi.; *Dickie in Journ. Linn. Soc.* xv. 44, 198; *Farlow in Bull. U. S. Nat. Mus.* 3, 30.

In positions still more open than *D. utilis.* (Cape Horn and the Falklands.)

1. **Desmarestia Rossii,** *Hook. f. and Harv., Flor. Antarct.* 467, t. clxxii., clxxiii.; *Dickie in Journ. Linn. Soc.* xv. 44, 198.

Swain's Bay on rocks in 3 fathoms, at the end of an island about 2 miles within the entrance of the bay, exposed to a slight swell from the open sea; local and not common, *Eaton.* (Fuegia, Falkland Islands, Heard Island, *Moseley.*)

2. **Desmarestia chordalis,** *Hook. f. and Harv., Flor. Antarct.* 467; *Dickie in Journ. Linn. Soc.* xv. 44, 198.

Swain's Bay, in 3 fathoms, with the preceding; very local, *Eaton.* Christmas Harbour, *Hooker, Moseley.* (Kerguelen Island only.)

A very graceful species. The fronds, upwards of 4 feet in length, are arranged in a manner similar to those of a fern, and cause the plant, as seen *in situ* from a boat, to bear a general resemblance in contour to such species as *Aspidium filix-mas.*

3. **Desmarestia aculeata,** *Lyngb.* Var. MEDIA, *Grev.; Hook. f. and Harv., Flor. Antarct.* 466; *Dickie in Journ. Linn. Soc.* xv. 44, 198.

Between tide-marks, Swain's Bay, *Eaton.* Cockburn Island (*Hooker*); and in North temperate and Arctic seas.)

4. **Desmarestia viridis,** *Lamx.;* D. VIRIDIS (and var. β. DISTANS), *Hook., Flor. Antarct.* ii. 466; *Dickie in Journ. Linn. Soc.* xv. 44, 198; *Farlow in Bull. U. S. Nat. Mus.* 3, 30.

Christmas Harbour, *Hooker, Moseley.* In 2 fathoms, Royal Sound and Swain's Bay; common, *Eaton.* (Marion Island; the Falklands; Cape Horn; American coast from New York northwards; Unalaschka; Hunde Island; W. coast of Europe.)

1. **Macrocystis pyrifera,** *Ag.; Flor. Antarct.* 461, t. clxix., clxx.; *Dickie in Journ. Linn. Soc.* xv. 44, 198; *Farlow in Bull. U. S. Nat. Mus.* 3, 30.

Abundant along rocky portions of the coast. (Antarctic Sea, from lat. 40° to 60° S.; New Zealand; Indian Ocean; Marion Island; Chili; California.)

1. **Lessonia fuscescens,** *Bory.; Flor. Antarct.* 457, t. clxvii., clxviii. A., and clxxi. D.; *Dickie in Journ. Linn. Soc.* xv. 44.

In exposed situations; Christmas Harbour, rare, *Hooker* and *Moseley*. (Chili, Fuegia, Falkland Islands, Cockburn Island, Auckland and Campbell Islands.)

1. **Asperococcus sinuosus,** *Roth; Flor. Antarct.* ii. 468; *Dickie in Journ. Linn. Soc.* xv. 198.

Crevices of rocks between tide-marks, Observatory Bay (two very small specimens), *Eaton*. (Widely distributed from the latitude of Spain to the Falklands; Florida; California; Japan.)

1. **Dictyosiphon fasciculatus,** *Hook. f. and Harv. Flor. Antarct.* 178, 167, t. lxix. 1.

Christmas Harbour, *Hooker*. (Falkland and Auckland Islands.)

1. **Adenocystis Lessonii,** *Hook. f. and Harv. Flor. Antarct.* i. 179, 468, t. lxix. 2 (details); *Dickie in Journ. Linn. Soc.* 1876, xv. 44, 198; *Farlow in Bull. U. S. Nat. Mus.* 3, 30.

Between tide-marks, abundant; scarcer but more finely grown in shallow estuaries (there, occasionally, as much as 5 or 6 inches long); Christmas Harbour, Royal Sound, Swain's Bay, &c. (Cape Horn; Falklands; Cockburn Island; Auckland and Campbell Islands.)

1. **Scytosiphon lomentarium,** *Grev.; Flor. Antarct.* 468 (*Chorda*).

Christmas Harbour, *Hooker*. (Falkland and Auckland Islands, Pacific Ocean to Japan and S. America; the Atlantic from the Faröe Islands to Cadiz; Mediterranean.)

1. **Elachista flaccida,** *Aresch.; Dickie in Journ. Linn. Soc.* xv. 199.

On *Rhodymenia palmata* in very shallow water along the shore in Observatory Bay, *Eaton*. (Atlantic coasts of France and Britain; Baffin's Bay.)

1. **Ectocarpus geminatus,** *Hook. f. and Harv. Flor. Antarct.* 469; *Dickie in Journ. Linn. Soc.* xv. 44, 199.

Plentiful on *Desmarestia* at Christmas Harbour, *Moseley*. Very slender solitary young plants on Annelid tubes, at 5 fathoms; stronger and more bushy, with trichosporangia only (but these abundant), on *Mytilus* at 1 fathom, and in tide pools, Observatory Bay; frequent in Swain's Bay among *Cladophora flexuosa* in pools and shallow water, *Eaton*. (Falklands and Cape Horn.)

1. **Sphacellaria corymbosa,** *Dickie in Journ. of Bot.* v. 50 (1876), *et in Journ. Linn. Soc.* xv. 199. ? S. funicularis, *Mont.; Flor. Antarct.* 469; *Farlow in Bull. U. S. Nat. Mus.*, 3, 30; fronde cæspitosa, filis cœspitosis, ramis inferne paucis dichotomis superne subpinnatim decompositis, ramulis alternis corymbosis.

On shells of *Mytilus* and on Annelid tubes; Swain's and Observatory Bays, *Eaton*.

The specimens are 2 to 3 inches long, cæspitose, but without fruit. Dr. Hooker's plant obtained at the Falklands, and Dr. Kidder's from the vicinage of Molloy Point in Royal Sound, are probably the same as the preceding.

2. **Sphacellaria affinis,** *Dickie in Journ. of Bot.* v. 50 (1876), *et in Journ.*

Linn. Soc. xv. 199; filis dense cæspitosis erectis parce dichotomis, articulis diametro subæqualibus vel paulo longioribus, trichosporangiis solitariis obovatis breviter pedicellatis.

On shells of *Mytitus* in rather open situations; Swain's Bay, *Eaton.*

The specimens are about ¼ inch in height, and are similar in habit to the British *S. radicaus.*

1. **Rhodomela Hookeriana,** *Ag.; Rhodomela Gaimardi Hook. f. and Harv. Flor. Antaret.* 481, t. clxxxiv. (*non Gaud.*); *Fartow in Butt.U. S. Nat. Mus.* 1876, 3, 30; *Dickie in Journ. Linn. Soc.* xv. 199.

Swain's and Observatory Bays, frequent, *Eaton;* near Molloy Point, one specimen, *Kidder.* (Falklands and Fuegia.)

1. **Polysiphonia abscissa,** *Hook. f. and Harv. Flor. Antaret.* 480, t. clxxxiii. 2; *Dickie in Journ. Linn. Soc.* xv. 199.

Forma microcarpa; P. microcarpa, *Hook. f. and Harv.* 479, t. clxxxii. 3; *Harv. Ner. Aust.* 42.

On roots of *Macrocystis,* and on tubes of Annelides, in 1 to 5 fathoms, Observatory Bay, *Eaton.* (New Zealand; Tasmania; Fuegia).

There are two forms of this species, one of them smaller and more rigid than the other. The *P. microcarpa* of the Flora Antarctica represents one of them, *P. abscissa* the other.

2. **Polysiphonia anisogona,** *Hook. f. and Harv. Flor. Antaret.* 478, t. clxxxii. 2; *Dickie in Journ. Linn. Soc.* xv. 44.

Kerguelen Island, *Mosetey.* (Falklands and Fuegia, *Hooker.*)

1. **Dasya Berkeleyi,** *Mont.; Dickie in Journ. Linn. Soc.* xv. 44, et 199 (var. β *Davisii*); *Fartow in Butt. U. S. Nat. Mus.* 3, 30. Polysiphonia punicea, *Hook. f. and Harv. Flor. Antaret.* i. 182; (Heterosiphonia) Berkeleyi? *var. β Davisii, idem,* 480.

Swain's Bay, on the seaward sides of islands, *Eaton;* Royal Sound, *Kidder.* (Auckland Islands, Marion Island, Falklands, Fuegia, Chilöe.)

Mr. Eaton's specimens belong to the var. β *Davisii,* having a habit and colour different from those of the typical plant. There are examples in different stages; but in all the ramuli are heterosiphonous.

1. **Bostrychia vaga,** *Hook. f. and Harv. Flor. Antaret.* 481, pl. clxxxvi. i. (Stictosiphonia).

Christmas Harbour, on rocks and stones above high water mark, and in damp places a considerable distance from the sea, abundant, *Hooker.*

1. **Delisea pulchra,** *Mont.; Flor. Antaret.* 484; *Dickie in Journ. Linn. Soc.* xv. 45.

Christmas Harbour, *Hooker, Mosetey.* (Heard Island, S. Tasmania, W. and E. Australia.)

1. **Ptilonia magellanica,** *Mont.; Dickie in Journ. Linn. Soc.* xv. 45, 200;

G

Farlow in Bull. U. S. Nat. Mus. 1876, 3, 30.   Plocamium ? magellanicum, *Hook. f. and Harv. Flor. Antarct.* 474.   Thamnophora magellanica, *Mont.*

Christmas Harbour, *Hooker, Moseley.*   In tideways and on parts of islands open to a slight swell from the outer sea, not in sheltered waters; Swain's Bay, *Eaton ;* Royal Sound, *Kidder.*   (Falklands and Fuegia.)

1. **Melobesia antarctica,** *Ag.; Dickie in Journ. Linn. Soc.* xv. 200.   M. verrucata, *Lamx; Dickie l. c.* 45.   M. verrucata, var. antarctica, *Hook. and Harv. Flor. Antarct.* 482.

On *Ballia, &c.,* Swain's Bay, *Eaton ;* Christmas Harbour, *Hooker.*   (Fuegia, Falklands, Tasmania, New Zealand, Auckland, Antarctic Seas.)

2. **Melobesia lichenoides,** *Ell. and Sol.; Dickie in Journ. Linn. Soc.* xv. 200.

Swain's Bay, common, *Eaton.*   The only example preserved was grappled in about 2 fathoms in a tideway between two islands, incrusting two sponges (*Microciona atrosanguinea,* Bk., and *Halichondria incrustans,* Jtn.; both British species.)   It is normal in habit, texture, and in the character of the keramidia, and is very luxuriant, measuring about 3 inches square.   (St. Paul's Island, Norfolk Island, the Mediterranean, France, Britain, Baffin's Bay.)

3. **Melobesia kerguelena,** *Dickie in Journ. Bot.* v. 51, 1876, *et in Journ. Linn. Soc.* xv. 200 ; simplex, 2½ poll. diam., dura, crassa, tantum in medio subtus adhærens, subconvexa, circumscriptione orbiculari, margine lævi parce undulata, keramidiis conspicuis numerosis plerumque in sericbus concentrice dispositis.

Swain's Bay in 2–3 fathoms, with the preceding, *Eaton.*

Mr. Eaton has an impression that this grows upon *Ballia* or *Ptilota,* but I should rather be disposed to suspect that it was attached to rocks.   The description was taken from an almost complete specimen; there are fragments of others whose contour is less regular, probably through interference of external objects.   All of them are in colour of a very pale buff or dull yellowish hue, varied with pale red tints.

1. **Delesseria Lyallii,** *Hook. f. and Harv. Flor. Antarct.* 471, t. clxxxvi; *Dickie in Journ. Linn. Soc.* xv. 45, 200 ; *Farlow in Bull. U. S. Nat. Mus.* 1876, 3, 30.

Christmas Harbour, *Hooker* (but not seen attached), *Moseley.*   Observatory Bay (ill grown), and Swain's Bay (well grown), abundant, *Eaton.*   (Marion Island, Falklands.)

Dr. Hooker obtaining only wrecked specimens at Christmas Harbour, was led to suppose that this species was a resident of the exposed coast.   It is however common in the very sheltered waters of Observatory Bay, though it certainly does not thrive there ; and it is abundant and luxuriant round the islands in Swain's Bay, in 3–5 fathoms.

2. **Delesseria Davisii,** *Hook. f and Harv., Flor. Antarct.* 470, t. clxxv.; *Dickie in Journ. Linn. Soc.* xv. 45, 200.

Swain's Bay; normal on *Mytilus* in sheltered places at or just below low-water mark; varying in more open situations, *Eaton.* (Falklands and Fuegia, *Hooker.*)

3. **Delesseria quercifolia,** *Bory.; Flor. Antarct.* 471; *Dickie in Journ. Linn. Soc.* xv. 200.

Swain's Bay, frequent, *Eaton.* (Falklands and Fuegia, *Hooker.*)

4. **Delesseria crassinervia,** *Mont.; Flor. Antarct.* 471; *Dickie in Journ. Linn. Soc.* xv. 200.

Swain's Bay, near the surface of the water; Observatory Bay, only one ill-grown example, *Eaton.* (Fuegia, Falkland, Auckland and Campbell Islands.)

1. **Nitophyllum fusco-rubrum,** *Hook. f. and Harv., Flor. Antarct.* 472; *Dickie in Journ. Linn. Soc.* xv. 45, 200; *Farlow in Bull. U. S. Nat. Mus.* 1876, 3, 30.

Christmas Harbour, *Hooker.* Abundant in open water in Swain's Bay, not found in sheltered places, *Eaton.* There is a variety in the collection with coccidia. Mouth of Royal Sound, *Kidder.*

2. **? Nitophyllum multinerve,** *Hook. f. and Harv., Flor. Antarct.* 473; *Dickie in Journ. Linn. Soc.* xv. 45.

Christmas Harbour (one specimen referred to this species with doubt, *Moseley*). (Falklands and Fuegia, *Hooker.*)

3. **Nitophyllum lividum,** *Hook. f. and Harv., Flor. Antarct.* 472, t. clxxix.; *Farlow in Bull. U. S. Nat. Mus.* 1876, 3, 30; *Dickie in Journ. Linn. Soc.* xv. 201.

In very sheltered water (one example only) at 6 to 10 fathoms, Swain's Bay, *Eaton.* Royal Sound, *Kidder.* (Falklands, *Hooker.*)

4. **Nitophyllum laciniatum,** *Hook. f. and Harv., in London Journ. of Bol.* iv. 256; *Dickie in Journ. Linn. Soc.* xv. 201. N. Bonnemaisoni *var.* laciniatum, *Hook. f. and Harv. l. c.* 474.

Swain's Bay, frequent, in 3 to 5 fathoms, *Eaton.* (Falklands and Fuegia, *Hooker.*)

1. **Chætangium variolosum,** *Mont.; Dickie in Journ. Linn. Soc.* xv. 45, 201. Notogenia variolosa, *Mont.; Flor. Antarct.* 487.

Christmas Harbour, very abundant, *Hooker, Moseley.* Observatory and Swain's Bays, abundant between tide marks. (Fuegia, Falklands, Auckland Islands.)

1. **Plocamium Hookeri,** *Harv. in Flor. Antarct.* 474; *Dickie in Journ. Linn. Soc.* xv. 45, 201.

Christmas Harbour, *Hooker, Moseley.* Swain's Bay, local, at 2 to 3 fathoms, in situations open to a slight swell from the outer sea, *Eaton.* (Heard Islands, *Moseley.*)

G 2

1. **Rhodophyllis capensis,** *Kütz.; Dickie in Journ. Linn. Soc.* xv. 201.

Swain's and Observatory Bays, sparingly, on tubes of Annelides, in 3–5 fathoms, *Eaton.* (Table and Simon's Bays.)

The few specimens collected by Mr. Eaton are dwarf and very narrow. They all have the structure of the genus, and must, I think, be referred to the above species.

1. **Rhodymenia palmata,** *Linn.; Flor. Antarct.* 475; *Farlow in Bull. U. S. Nat. Mus.* 1876, 3, 30; *Dickie in Journ. Linn. Soc.* xv. 201.

Swain's Bay and Royal Sound, common in tide pools and shallow water; very luxuriant specimens. Also a dwarf form of olivaceous complexion, growing between tide marks, dry at low water, in Swain's Bay, *Eaton.* (Falkland Islands; Fuegia; Unalaschka; Greenland; Newfoundland; Scandinavian, British, and French coasts.)

2. **Rhodymenia corallina,** *Grev.; Flor. Antarct.* 475; *Farlow in Bull. U. S. Nat. Mus.* 1876, 3, 30.

On roots of *Macrocystis,* Christmas Harbour, rare, *Hooker;* Royal Sound, *Kidder.*

1. **Phyllophora cuneifolia,** *Hook. f. and Harv., Flor. Antarct.* 486; *Dickie in Journ. Linn. Soc.* xv. 201. P. Brodiæi, *Turn.? Flor. Antarct. l. c.; Dickie l. c.*

Christmas Harbour, rare, *Hooker.* Swain's Bay, rare, in very sheltered water, at 5–10 fathoms, *Eaton.* (Falkland Islands.)

Professor Agardh (loc. cit.) considers with the authors of this species that it is probably a form of *P. Brodiæi.*

1. **Ahnfeltia plicata,** *Huds.; Dickie in Journ. Linn. Soc.* xv. 46, 201. Gigartina plicata, *Hook. f. and Harv., Flor. Antarct.* 487.

Local, between tide marks near Observatory Bay, *Eaton;* Christmas Harbour, abundant, *Hooker.* (Falkland Islands; temperate and colder seas in the northern hemisphere.)

1. **Callophyllis variegata,** *Bory.; Dickie in Journ. Linn. Soc.* xv. 46, 201; *Farlow in Bull. U. S. Nat. Mus.* 1876, 3, 31. Rhodymenia variegata (*in part*), *Hook. f. and Harv. Flor. Antarct.* 475.

Christmas Harbour, *Hooker.* Swain's and Observatory Bays, in sheltered situations, *Eaton.* Royal Sound, *Kidder.* (Auckland Islands; New Guinea; Falklands; Fuegia; Chili; Peru; California.)

Mr. Eaton's collection comprises different forms of this very variable species :— from Observatory Bay, on *Mytilus* in sheltered water, a variety with small marginal kalidia; from Swain's Bay var. β *atro-sanguinea,* also a narrow variety (?), torn at the apex and proliferous; and in addition var. γ on roots of *Macrocystis* in very sheltered water.

2. **Callophyllis dichotoma,** *Kütz.* Rhodomenia dichotoma, *Hook. f. and Harv., Ftor. Antarct.* 186, t. lxxii. 1.

Swain's Bay, one specimen only, *Eaton.* (Marion Island, *Moseley;* Campbell Island, *Hooker.*)

The specimen obtained at Kerguelen Island by Mr. Eaton has the structure and kalidia of *Cattophyllis.* The last are not marginal, and therefore it is not a form of *C. variegata.* [This species was not mentioned in Dr. Dickie's MS., nor in his list in the Linnæan Society's Journal; but the name and remarks were noted by him on the sheet containing the specimen in the collection, A. E. E.]

3. **Callophyllis tenera,** *J. Ag.; Dickie in Journ. Linn. Soc.* xv. 202.

Local in very sheltered water, Swain's Bay, *Eaton.* (South Shetlands.)

1. **Kallymenia dentata,** *Suhr.* (Halymenia), *Dickie in Journ. Linn. Soc.* xv. 16 (*vars. a and* γ), 202.

Swain's Bay; and (at 1 fathom, of inferior growth) Observatory Bay, *Eaton.* (Cape of Good Hope.)

1. **Gigartina Radula,** *Esp.; Dickie in Journ. Linn. Soc.* xv. 46, 202; *Farlow in Butt. U. S. Nat. Mus.* 1876, 3, 30. Iridæa Radula, *Hook. f. and Harv., Ftor. Antarct.* 485.

Christmas Harbour, *Hooker.* Swain's and Observatory Bays, abundant on rocks from low-water mark to 1 fathom or more, *Eaton.* (Cape of Good Hope; New Zealand; Auckland and Campbell Islands; California.)

The collection includes various forms of this species.

1. **Iridæa capensis,** *J. Ag.; Dickie in Journ. Linn. Soc.* xv. 46.

Kerguelen Island, *Moseley.* (Cape of Good Hope.)

2. **Iridæa laminarioides,** *Bory.; Dickie in Journ. Linn. Soc.* xv. 46.

Kerguelen Island, *Moseley.* (Auckland Islands, and the S.W. shores of Chili.)

Mr. Moseley's collection contains several specimens which belong, I think, to this species.

1. **Epymenia variolosa,** *Kütz.; Dickie in Journ. Linn. Soc.* xv. 45. Rhodymenia variolosa, *Hook. f. and Harv., Ftor. Antarct.* 476, clxxx.; *Dickie l. c.; Farlow in Butl. U. S. Nat. Mus.* 3, 30.

Christmas Harbour, *Hooker.* Swain's Bay, local, *Eaton.* Royal Sound, *Kidder.*

1. **Halymenia latissima,** *Hook. f. and Harv., Ftor. Antarct.* 189, t. lxxiii., 1, 2; *Dickie in Journ. Linn. Soc.* xv. 202.

Swain's and Observatory Bays; common on rocks at low-water mark, and on *Mytilus* at 1 fathom, *Eaton.* (Auckland and Campbell Islands, *Hooker.*)

1. **Ceramium rubrum,** *Ag.; Ftor. Antarct.* 488; *Dickie in Journ. Linn. Soc.* xv. 46, 202. C. rubrum var. secundatum, *Lyngb.; Farlow in Bull. U. S. Nat. Mus.* 3, 31.

Christmas Harbour, very abundant, *Hooker.* Common in Swain's Bay, Royal Sound, and near Vulcan Cove, *Eaton.* (General in the colder seas of both hemispheres.)

2. **Ceramium diaphanum,** *J. Ag.; Flor. Antarct.* 488.

Christmas Harbour, abundant, *Hooker.* (Cape of Good Hope and Atlantic coasts of Europe.)

1. **Ptilota Eatoni,** *Dickie in Journ. of Bot.* v. 51, 1876; *et in Journ. Linn. Soc.* xv. 202.; rachide filiformi 5-6-pollicari, pinnis oppositis inæqualibus, unâ majore alternâ minore, pinnulis pinnarum longiorum apices versus pectinatis, reliquiis bases harum versus, cæterisque omnibus subulatis ex serie articulorum magnorum subsimplici structis, sphærosporis ad apices pinnularum subsolitariis, favellis terminalibus, ramis involucri 4-5 pectinatis conniventibus.

Swain's Bay, in 2 to 5 fathoms, on the side and end of a promontory and of an island about two miles from the entrance of the bay, exposed to the tide and a slight swell from the outer sea; very local. Usually parasitic upon *Battia*, sometimes attached to *Mytilus; Eaton.*

This species resembles *P. Harveyi* in the character of the involucre, but differs from it in general habit, and in the structure of the larger and smaller pinnules. It is also dissimilar in colour, being dull purple.

PLATE V., Fig. iii. :—1, portion of frond of *nat. size ;* 2, portion of stem with young branch ; 3, apex of fully grown branch ; 4, ditto with sphærospores ; 5, sphærospores ; all much *entarged.*

1. **Ballia callitricha,** *Ag.; Dickie in Journ. Linn. Soc.* xv. 46, 202 ; *Farlow in Bull. U. S. Nat. Mus.* 1876, 3, 31. Ballia Brunonis *var.* β Hombroniana, *Hook. f. and Harv., Flor. Antarct.* 488.

On *Mytilus,* roots of *Macrocystis,* and Annelid tubes, from tide pools down to 6 fathoms ; very common in Christmas Harbour, Swain's Bay, and Royal Sound. (Falklands ; Marion Island ; Australia ; Tasmania ; New Zealand ; Auckland Islands.)

1. **Callithamnion simile,** *Hook.f. and Harv., Flor. Antarct.* 489; *Dickie in Journ. Linn. Soc.* xv. 202.

Christmas Harbour, rare, *Hooker.* On *Mytilus,* Annelid tubes, and roots of *Macrocystis,* in 1 to 5 fathoms, in Observatory and Swain's Bays; frequent, *Eaton.*

2. **Callithamnion Ptilota,** *Hook. f. and Harv., Flor. Antarct.* 489, t. clxxxix. 1; *Farlow. in Bull. U. S. Nat. Mus.* 3, 31.

Royal Sound, *Kidder.* (Crozets, *Hooker.*)

3. **Callithamnion Rothii,** *Lyngb.; Dickie in Journ. Linn. Soc.* xv. 203.

In tide pools and at the extreme verge of low water, on *Mytilus,* in Swain's and Observatory Bays, local, *Eaton.* (Atlantic shores from Greenland to Africa ; N.E. shores of the United States.)

I can see no essential difference between Mr. Eaton's specimens and the plant from the northern hemisphere. They agree in habit, and in the arrangement of the tetraspores. The articulations are a little longer than those of British examples.

1. **Codium adhærens,** *Ag.; Farlow in Bull. U. S. Nat. Mus.* 1876, 3, 31; *Dickie in Journ. Linn. Soc.* xv. 203.

On rocks in about 2 fathoms in Observatory Bay; frequent, *Eaton.* (Europe; Cape of Good Hope; Mauritius; Ceylon; Australia; Friendly and Loo-choo Islands.)

2. **Codium tomentosum,** *Stackh.; Flor. Antarct.* 491; *Dickie in Journ. Linn. Soc.* xv., 46.

Christmas Harbour, *Hooker.* (Tongabu; Banda Islands; and the colder seas of both hemispheres).

1. **Bryopsis plumosa,** *Grev.; Flor. Antarct.* 492; *Dickie in Journ. Linn. Soc.* xv. 203.

Dwarf or very young specimens on Annelid tubes in 5 fathoms, Observatory Bay, scarce, *Eaton.* (Greenland; widely distributed throughout both the temperate zones, and even in some of the warmer seas).

1. **Vaucheria Dillwynii,** *Ag.; Flor. Antarct.* ii. 492.

On the ground amongst Penguin rookeries, Christmas Harbour, *Hooker.*

1. **Ulva latissima,** *Linn.; Flor. Antarct.* 499; *Dickie in Journ. Linn. Soc.* xv. 47, 203; *Farlow in Bull. U. S. Nat. Mus.,* 1876, 3, 31.

Christmas Harbour, very common, *Hooker.* Royal Sound; Swain's Bay, *Eaton.* (Widely distributed in both hemispheres.)

2. **Ulva (?) cristata,** *Hook. f. and Harc.; Flor. Antarct.* 499.

In moist clefts of rocks overhanging Christmas Harbour, growing with *Tryptholallus* [*Palmodictyon,* Ktz.] *anastomosans, Hooker.*

1. **Porphyra laciniata,** *Ag.; Flor. Antarct.* 500; *Dickie in Journ. Linn. Soc.* xv. 46, 203.

Christmas Harbour, very abundant, *Hooker.* Common in shallow water, Observatory Bay. (Temperate and colder seas of both hemispheres.)

2. **Porphyra vulgaris,** *Ag.; Flor. Antarct.* ii. 500; *Dickie in Journ. Linn. Soc.* xv. 46.

Christmas Harbour, very abundant, *Hooker.* (Widely distributed in the northern and southern hemispheres.)

1. **Enteromorpha compressa,** *Link.; Flor. Antarct.* 500; *Dickie in Journ. Linn. Soc.* xv. 46, 203.

Very abundant on rocks and in tide-pools; Christmas Harbour, *Hooker;* Royal Sound and Swain's Bay, *Eaton.* (Cosmopolitan).

2. **Enteromorpha intestinalis,** *Link.; Flor. Antarct.* 500.

Christmas Harbour, *Hooker.* (Cosmopolitan.)

1. **Prasiola fluviatilis**, *Sommerfeldt, Supplem. Fl. Lapp.* 1826 (*teste Dickie in Arctic Manual*, 1876.) P. Sauteri, *Meneghini MS., Kütz. Sp. Alg.* 1849, p. 473; *Dickie in Journ. Linn. Soc.* xv. 203.

On wet rocks in the bed of a streamlet descending from a patch of snow, about 400 feet above the sea, on the pyramidal hill on the W. side of Swain's Bay. (European Alps to 9,300 ft.; Spitsbergen; streams of W. Greenland.)

1. **Cladophora rupestris**, *Linn.; Ftor. Antaret.* 495; *Dickie in Journ. Linn. Soc.* xv. 47.

Christmas Harbour, on rocks, *Hooker*. (General between the arctic circle and the Mediterranean; only at Kerguelen Island in the southern hemisphere).

2. **Cladophora arcta**, *Ktz.; Ftor. Antaret.* 495; *Dickie in Journ. Linn. Soc.* xv. 47, 203. C. Hookeriana, *Ktz. Sp. Alg.*, p. 418.

Very abundant on *Mytilus* at low-water mark, and in shallow water, Observatory Bay, *Eaton*. (Falkland Islands; Fuegia; German and N. Atlantic Oceans; Greenland.)

3. **Cladophora simpliciuscula**, *Hook. f. and Harv. Ftor. Ant.* 406, t. cxcii. 4, 1–3; *Dickie in Journ. Linn. Soc.* xv. 203.

One very small specimen, probably very young, too imperfect to be identified with absolute certainty, but which I think may be referred to this species, was obtained on an Annelid tube in Observatory Bay at 5 fathoms, *Eaton*. (Falklands and Fuegia, *Hooker*.)

4. **Cladophora flexuosa**, *Griff; Dickie in Journ. Linn. Soc.* xv. 203.

In tide-pools and at 5 fathoms in Observatory and Swain's Bays, specimens from the deeper water are poor, *Eaton*. (Shores of Europe; Massachusetts Bay.)

1. **Rhizoclonium riparium**, *Roth.; Cladophora riparia, Hook. f. and Harv. Ftor. Antaret.* 495.

Christmas Harbour, *Hooker*. (Cumberland Sound; British coasts, &c.)

2. **Rhizoclonium ambiguum**, *Ktz.; Conferva ambigua, Hook. f. and Harv. Ftor. Antaret.* 494, t. cxci. 1.

Christmas Harbour, in the sea, *Hooker*.

1. **Chætomorpha Linum**, *Ktz.; Conferva Linum, Roth.; Ftor. Antaret.* 493.

Christmas Harbour, on rocks near high-water mark, *Hooker*.

VI.—*Fresh-water Algæ collected by the Rev. A. E. Eaton.*

Algæ aquæ dulcis Insulæ Kerguelensis,

auctore

PAULO FRIDERICO REINSCH.

(Cum notulis de distributione geographica a G. Dickie adjectis.)

Tanto ampliores notitiæ de plantis simplicissima structura ac simplicissimis organis in terris diversissimis crescentibus, quo evidentius factum agnoscitur notandum : has plantas iisdem legibus non subjectas ex quibus dependet plantarum in systemate superiorum diffusio in orbi. Specierum plantarum microscopicarum diffusio universa determinatus rationibus peculiaribus : diffusione aeris meatus in superficie terræ effecta, mobilitate levissima cellularum propagativarum earumque vi vitali diu permanente in statu ipso siccato; neque minus efficitur diffusio rationibus vitæ multo simplicioribus accommodatis ad diversissima terræ cœla.

Materia hujus enumerationis algarum Insulæ selecta a Cl. A. Eaton in expeditione transitus Veneris in hieme 1874–5 continet numeros quatuordecim. Speciminum inquirendi causa ab Herbario Regio Kewensi mihi traditorum sunt : ampullulæ tres cum algis aquæ dulcis in spiritu vini asservatis, decem folia cum algis siccatis et capsula parvula cum algis siccatis. Omnia quæ ampullula major continuerat : Specimina compluria muscorum aquatilium densissime algis variis (*Schizosiphon* Spec. nova et *Nostoc* Spec. var.) vestita cum singulis speciminibus *Nitellæ antarticæ* et *Vaucheriarum* cæspitulis parvulis intermixta : mihi dedit materiam pro maximam partem hujus enumerationis. In hac ampullula inventæ erant 81 species algarum aquæ dulcis ad 45 genera spectantes; harum algarum sunt novæ species 28, nova genera 3.

Omnes in hac enumeratione receptæ species sunt conjunctæ in præparatorum collectione integra, nunc in Herbario Regio Kewensi deposita.

Insulæ Kerguelensis Specierum algarum aquæ dulcis hucusque cognitarum numerus totus est 106, numerus generum 67.

Ab his sunt

Diatomophyceæ	- - -	21 species, 13 genera.
Phycochromophyceæ	- -	33 species, 18 genera.
Chlorophyllophyceæ ·	-	50 species, 31 genera.
Melanophyceæ et Rhodophyceæ	-	2 species, 2 genera.

Omnes Familiæ Algarun aquæ dulcis, *Ulothrichaceis, Chroolepideis, Confercaceis, Sphæropleaceis* exceptis, inveniuntur in Insula Kerguelensi. A *Cladophoris, Chætophoris, Rhizocloniis* nulla species est observata. In ordinem systematis adducta Flora Algarum aquæ dulcis Insulæ hucusque cognita, est hæc.

** II

DIATOMOPHYCEÆ.

13 genera, 21 species. (2 Spec. novæ, 3 Spec. incert.)

PHYCOCHROMOPHYCEÆ.

Chroococcaceæ.—5 genera, 6 species (ab his 1 novæ, 1 incerta).
Oscillariaceæ.—3 genera, 3 species (ab his 2 novæ, 1 forma nova).
Nostochaceæ.—3 genera, 10 species (ab his 7 novæ, 1 forma nova).
Rivulariaceæ.—3 genera, 5 species (ab his 2 novæ, 2 formæ novæ).
Scytonemaceæ.—3 genera, 4 species (ab his 1 nova, 2 formæ novæ).
Sirosiphoniaceæ.—1 genus, 5 species (ab his 4 novæ, 1 forma nova).

CHLOROPHYLLOPHYCEÆ.

Palmellaceæ.—7 genera, 9 species (2 novæ formæ).
Protococcaceæ.—4 genera, 5 species (1 genus novum, 1 species nova).
Volvocineæ.—1 genus, 1 species. (Spec. nova?)
Desmidieæ.—4 genera, 5 species (1 nova, 3 formæ novæ).
Zygnemeæ.—4 genera, 7 species (1 nova, 1 forma nova).
Vaucheriaceæ.—3 genera, 6 species (2 novæ).
Ulvaceæ.—1 genus, 1 species.
Oedogoniaceæ.—2 genera, 5 species (2 species incertæ).
Chætophoraceæ.—7 genera, 10 species.
    a. Chætophoreæ.—4 genera, 6 species (1 genus novum, 5 species novæ, 1 forma nova).
    b. Gongrosireæ.—3 genera, 4 species (1 nova).

RHODOPHYCEÆ.

1 genus, 1 species nova.

MELANOPHYCEÆ.

1 genus novum, 1 species nova.

DIATOMOPHYCEÆ.*

1. **Stauroneis goeppertiana**, *Bleisch; Rabenhorst Alg. Europ. Nr.* 182; specimina kerguelenensia accuratissime consentiunt in magnitudine ac forma cellularum cum speciminibus Silesiacis in Collect. Algar. Rabenhorst. communicatis. Arcola transversalis in speciminibus Silesiacis plurimis paulo est angus-

---

* Materiam enumerationis Diatomacearum in ampullula majore reperi in singulis terrosis corpusculis duris radiculis *Nitellæ antarcticæ* partim adhærentibus partim in liquore fluitantibus.

tior,—Longit. 0,0224 mm. ($\frac{1}{9^1_4}'''$ Engl.) Latit. 0,0058 mm. ($\frac{1}{3^1_3^4}'''$ Engl.)—(DISTRIB. Silesia.—Considered by some authorities to be probably a form of *S. dilatata*, which is widely distributed in Europe, *G. Dickie*.)

2. **Stauroneis anceps,** *Ehrenberg ;* forma linearis. Maxime consentiunt specimina cum speciminibus Europæis a Erlangen et a Falaise leg. Brébisson.— (DISTRIB. Europa tota, California, Cayenne.)

3. **Stauroneis Phœnicenteron,** *Ehrenberg.*—Longit. 0,0952 mm. ($\frac{1}{2^1_2}'''$ Engl.)—(DISTRIB. Europa frequens, America, Persia.)

1. **Aclinanthes exilis,** *Kützing.* Longit. 0,0224 mm. ($\frac{1}{9^1_4}'''$ Engl.) Latit. 0,0028 mm. ($\frac{1}{7^1_5^7}'''$ Engl.) In quoque latere in medio cellulæ (a fronte visæ) nodulus singulus ; in plurimis speciminibus Europæis tantum in interiore latere. In magnitudine ac forma maxime consentiunt cum speciminibus e Jura Franconia, e Hungaria, et e Falaise (Gallia).

In *Vaucheriæ* cellulis nidulans.—(DISTRIB. in Europa vulgaris.)

1. **Larirella diaphana,** *Rcisch.* Longit. 0,1008 mm. ($\frac{1}{3^1_4}'''$ Engl.) Latit. 0,0418 mm. ($\frac{1}{5^1_0}'''$ Engl.) In speciminibus singulis.—(DISTRIB. Silesia ; an forma *S. splendidæ* in Europa vulgaris, *G. Dickie*.)

1. **Campylodiscus,** *species nova ; Rcinsch. in. Journ. Linn. Soc.* xv. 205 ; magnus, elliptico-ovalis, utroque polo rotundato-obtuso, costis marginalibus radialibus crassis usque ad tertiam partem latitudinis superficieis se pertinentibus in quoque latere 22is—24is, areolas 21as—22as rectangulares includentibus, area media lævi ; frustulæ a latere visæ simpliciter spiraliter curvatæ areolis 21is—22is rectangularibus instructæ.—Diam. longit. 0,132 mm. ($\frac{1}{3^1_6}'''$ Engl.)—Diam. transvers. 0,0666 mm. ($\frac{1}{3^1}$ Engl.) — Costæ in 0,02 mm. ($\frac{1}{11^1_6}'''$ Engl.) tres.

In speciminibus singulis inter *Schizosiphontis kerguelensis* cæspites.

A *Campylodiscis* frustulis oblongis *Campylod. Larirella*, Ehrenberg, (Abh. Berlin. Acad. 1845, p. 362), mihi tantum ex diagnosi nota, proxima species videtur.

1. **Gomphonema Brebissonii,** *Kützing, Spec. Alg.* p. 66 ; *Ralfs Brit. Infus.* p. 887. Gomph. acuminatum var. *Smith. Brit. Diat.*—Longit. 0,0478 mm. ($\frac{1}{1^1_4}'''$ Engl.) Latit. (in summo) 0,0112 mm. ($\frac{1}{8^1_9}'''$ Engl.)—Cum speciminibus e locis diversis Germaniæ et Austriæ et e Falaise Gallia maxime consentiunt.

In *Vaucheriæ sessilis* et sericeæ filis.—(DISTRIB. in Europa vulgaris, an forma *G. dichotomi? G. Dickie*.)

1. **Amphiprora Spec. nova,** *Reinsch in Journ. Linn. Soc.* xv. 205 ; parva, rectangularis, subtilissime striata, medio parum constricta, utroque polo late truncato-rotundato, lineis intermediis duabus in medio cellulæ æqualiter extrorsum curvatis aream mediam cruciformem lævem circumcingentibus, nodulo singulo et in quoque latere cellulæ in decussi linearum incluso et in summo utriusque lineæ. Longit. 0,0333 mm. ($\frac{1}{c^1_3}'''$ Engl.) Latit. 0,0084 mm. ($\frac{1}{2^1_4^3}'''$ Engl.)

Amph. Pockorugana Grunow : dimensionibus duplo majoribus cellulis ovato-

u 2

oblongis rotuudatis, nodulo centrali, *Amph. minor* Gregory : dimensionibus majoribus cellulis elliptico-oblongis polis rotundatis, striis radiatis differunt.

1. **Navicula elliptica,** *Kützing,* var. Cocconeides ; *Rabenhorst, Algenfl. Europ.* I., p. 180, dimensionibus duplo minoribus. Longit. 0,0201—0,0224 mm. ($\frac{1}{50}$ —$\frac{1}{44}$''' Engl.) Latit. 0,013—0,0168 mm. ($\frac{1}{137}$—$\frac{1}{128}$''' Engl.)

In opere novissimo de Diatomaceis, Atlas der Diatomaceenkunde, Heft II. tab. VII., fig. 55, *Navicula* est delineata (e Germania) quae maxime consentit in forma, magnitudine, ac structura cellulae cum plantula Kerguelensis.—(Distrib. in Europa frequens, Java, Nova Zelandia.)

2. **Navicula dicephala,** *Ehrenberg.* Longit. 0,0248 mm. ($\frac{1}{47}$''' Engl.) Speciminа ad formam pertinent summis capituliformibus distinctius disjunctis. —(Distrib. Europa).

3. **Navicula minutissima,** *Grunow.* E. minimis cellularis lineari-oblongis, nodulo medio et linea longitudinali distincta, indistincte transversaliter striatis.— Longit. 0,0112 mm. ($\frac{1}{150}$''' Engl.) Latit. 0,0028 mm. ($\frac{1}{517}$'' Engl.)

Hæc *Navicula* in speciminibus numerosis in massa ex Diatomaceis exstituta ; non sine dubio *Navicula kerguelensis* ad hauc speciem est posito.— (Distrib. Europa orientalis.)

4. **Naviculæ Spec.** Cellulis lanceolatis apicibus capituliformibus porrectis, nodulo centrali et linea media indistincta, marginibus distincte striatis striis ad mediam non pertinentibus. Longit. 0,0278 mm. ($\frac{1}{75}$''' Engl.) Latit. 0,0056 mm. ($\frac{1}{375}$''' Engl.)

1. **Amphora gracilis,** *Ehrenberg.* Longit. 0,0357 mm. ($\frac{1}{58}$''' Engl.) Latit. 0,0196 mm. ($\frac{1}{107}$ Engl.) Cellulæ ovato-ovales apicibus truncatis, nodulis circa tertiam partem diametri transversalis a margine distautibus, areola media subtiliter longitudinaliter striato. Specimina a Falaise (Gallia) et e Germania consentiunt in magnitudine ac forma cellularum. *Amphoræ gracilis* in Schmidt Atlas d. Diatomaceenkunde, vii. fasc. tab. 26, fig. 101, cellulæ, quæ ad *Amphoram angustam* Ehrenberg pertineut, graciliores et apicibus attenuatis.—(Distrib. Europa, Mexico, et in Kurdistania fossilis.)

1. **Pinnularia viridula,** *Smith Brit. Diatom.* 57, tab. xviii. fig. 179; *Rabenh. Eur. Alg.* i. p. 214. Forma apicibus subito attenuatis, striis transversalibus lineam mediam attingentibus distinctissimis. Longit. 0,0357mm. ($\frac{1}{58}$''' Engl.) Latit. 0,0123mm. ($\frac{1}{168}$''' Engl.)

Plantulæ Falaise (A. de Brebisson, leg.) et plantulæ Erlangensis in magnitudine consentiunt, sed differunt polis sensim attenuatis striis subtilioribus. (Distrib. Europa, America.)

2. **Pinnularia viridis,** *Ehrenberg.* Longit. 0,0648mm. ($\frac{1}{32}$''' Engl.) Latit. 0,013mm. ($\frac{1}{157}$''' Engl.) Specimina paulo minora speciminibus formæ apud Erlaugam communis. (Distrib. Europa, America, Persia.)

3. **Pinnulariæ** species; cellulis ovato-ellipticis, polis rotundatis, nodulo cen-

trali firmo, striis transversalibus distinctis lineam mediam attingentibus. Longit. 0,0168—0,0196mm. $(_1\frac{1}{6}—_1\frac{1}{9}''$ Engl.) Latit. 0,0084mm. $(_2\frac{1}{5\,8}'''$ Engl.)

4. **Pinnulariæ** species; cellulis late ovato-ellipticis, polis subito angustatis apicibus rotundatis, nodulo centrali firmo striis transversalibus distinctis lineam mediam attingentibus. Longit. 0,0224mm. $(_0\frac{1}{5}'''$ Engl.) Latit. 0,0112mm. $(_1\frac{1}{6\,0}'''$ Engl.)

1. **Synedra Vaucheriæ**, *Kützing.*; forma apicibus obtusis. Longit. 0,0448mm. $(\frac{1}{7}'''$ Engl.) Latit. 0,0028—0,0056mm. $(_7\frac{1}{2\,7}—_3\frac{1}{5\,6}''$ Engl.)

Individua breviter stipitata basi radiatim conjuncta in *Schizosiphonte kerguelensi*, et in *Vaucheriæ* cellulis. (DISTRIB. in Europa frequens.)

1. **Eunotia pectinalis**, *Dillwyn.* Longit. 0,106mm. $(_2\frac{1}{0}'''$ Engl.). Latit. 0,0393mm. $(_5\frac{1}{4}'''$ Engl.) (DISTRIB. in Europa vulgaris.)

1. **Denticula thermalis**, *Kützing.*, var. minor. Longit. 0,0168mm. $(_1\frac{1}{2\,5}'''$ Engl.) Latit. 0,0056mm. $(_3\frac{1}{5}'''$ Engl.) Cellulæ in quoque latere 9is nodulis instructæ. (DISTRIB. Aquis calidis Galliæ, Hungariæ, et Italiæ.)

1. **Cymbella gastroides**, *Ehrenberg.* Longit. 0,0421mm. $(_5\frac{1}{6}'''$ Engl.) Latit. 0,013mm. $(_1\frac{1}{3\,7}'''$ Engl.)

Specimina minora speciminibus e locis variis Germaniæ. (DISTRIB. Per totam Europam.)

PHYCOCHROMOPHYCEÆ.

1. **Chroococcus macrococcus**, *Rabenh.*, *Alg. Fl. Eur.* i. p. 33. Protococcus macrococcus, *Kütz.*, *Tab. Phyc.* i. tab. 2. Forma cytiodermate tenuiore, cytioplasmate grossius granuloso.

Formam typicam tantummodo observavi in familia singula tricellulari inter algas unicellulares *Hormosiphonti* adhærentes. Diam. cellular. indivis. 0,0478mm. $(_4\frac{1}{0}'''$ Engl.)

Formam in familiis singulis bi- et tricellularibus in massis minoribus algarum vaiiarum foliis muscorum adhærentibus observavi. Hæc forma pertinet ad formam *Chr. macrococci* = *Chroococcus aureus*, Kütz., Tab. Phyc. ii. tab. 2, Chrooc. macrococc. Rabenh., var. β.; cytioplasmatis cellularum colorem nunc pallide-flavum in statu vivente cellulæ fuisse aureo-luteum non dubito. (DISTRIB. Europa tota.)

1. **Microcystis olivacea**, *Kütz.*, *Tab. Phyc.* i. tab. 9. Diam. cellular. 0,0041mm. $(_5\frac{1}{6\,7}'''$ Engl.) Diam. famil. 0,066—0,0896mm. $(_3\frac{1}{1}—_2\frac{1}{3}'''$ Engl.)

Observavi tantum familias singulas inter alias algas unicellulares muscis adhærentes. In colore quoque obscure olivaceo cum specim. Europæis consentit. (DISTRIB. Germania.)

2. **Microcystis parasitica**, *Kütz.*, *Tab. Phyc.* i. tab. 9, fig. 1.

In physcumatum structura, magnitudine ac colore cellularum cum speciminibus Europæis et cum icone Kützingiana maxime consentiens. Physcumata minora et majora, partim cohærentia in *Nitellæ antarcticæ* cellulis affixa.

Diam. cellular. 0,003mm. ($\frac{1}{744}'''$ Engl.) Diam. physeumatum 0,0278—0,112mm. ($\frac{1}{45}$—$\frac{1}{9}'''$ Engl.) (DISTRIB. Europa.)

1. **Glœthece involuta**, *Reinsch. in Journ. Linn. Soc.* xv. 206; thallo non limitato inter algas minores disperso; cellulis oblongo-cylindricis utroque polo rotundatis, diametro transversali duplo longioribus, singulis aut geminis, tegumentis binis crassissimis distincte plurilamellosis circumvelatis, cytioplasmate pallide-æruginco subtiliter granuloso, plerumque granulo singulo majore instructo. Longit. cellular. (c. indum. exter.) 0,0278—0,0333mm. ($\frac{1}{715}$—$\frac{1}{93}'''$ Engl.) Longit. cellular. (c. indum. inter.) 0,0248—0,0278mm. ($\frac{1}{814}$—$\frac{1}{715}'''$ Engl.) Longit. cellul. (sine indum.) 0,0112—0,013mm. ($\frac{1}{719}$—$\frac{1}{737}'''$ Engl.)

Inter algas minores muscis aquaticis insidentes. Ilæc *Glœthecc* aliqua similitudine consentit cum *G. monococca*, Rabenh, Fl. Eur. i. p. 62=*Glœcapsa monococca*, Kütz., Tab. Phyc. i. tab. 23, itidem reperta plerumque in statu uni- et bicellulari; quæ species nova attamen est diversa indumenti structura valde distincte lamellosa et cellulis longioribus a *Glœth. monococca*, cujus integumentum semper est structura homogenea.

2. **Glœcapsa magna**, *Kütz., Tab. Phyc.* i. tab. 22, fig. 7.

Cellulæ singulæ et geminæ sphæricæ colore intensive ærugineo-viridi densissime positæ, physeumata sphærica plerumque cohærentia indumento colorato velata formantes. Cellular. diam. 0,028—0,0041mm. ($\frac{1}{737}$—$\frac{1}{547}'''$ Engl.) Diam. physeumatis 0,033—0,05mm. ($\frac{1}{715}$—$\frac{1}{42}'''$ Engl.)

Inter *Scytonema castaneum*, Kütz., in massis parvulis *Hormosiphonti coriacco* adhærentibus " prope Vulcan Cove." Non est mihi certissime, hanc plantulam pertinere ad *Gl. magnam* propter speciminum observatorum minimum numerum. (DISTRIB. Europa, Greenlandia.)

1. **Anacystis marginata**, *Meneghini.*

Familiæ singulæ quarum diameter 0,17mm. ($\frac{1}{6}'''$ Engl.), inter Algarum massas minores natantes. (DISTRIB. Europa.)

1. **Leptothrix hyalina**, *Reinsch in Journ. Linn. Soc.* xv. 206; aggregata, cæspitulos dispersos et radicantes muscis aquaticis affixas formans, trichomatibus hyaliuis, vaginis distinctissimis crassis hyalinis, superne sæpissime vacuis et in summo apertis, cellulis tenuissimis diametro æqualibus, cytioplasmate punctulato. Diam. trichomat. 0,0028—0,0041mm. ($\frac{1}{737}$—$\frac{1}{547}'''$ Engl.) Cæspitulorum altitudo, 0,084—0,112mm. ($\frac{1}{25}$—$\frac{1}{9}'''$ Engl.)

In foliis muscorum.

Leptothrix radians, Kütz., Tab. Phyc. ii. tab. 59, proxima species distinguitur vaginis multo angustioribus cellulis crassioribus.

1. **Lyngbya major**, *Kütz., Tab. Phyc.* i. *tab.* 90, *fig.* 8; var. *kerguelenensis;* trichomatibus inter alias algas dispersis subrectis, cellulis intensive ærugineis subtiliter distincte granulatis, diametro 8plo—10plo brevioribus, vaginis amplis hyalinis (interdum fuscescentibus) distincte 8–12—lamellosis, cellulis interstialibus

nullis. Diam. trichomat. (c. vagin.), 0,0361—0,0448mm. $(_d{}^1_0—_1^1{}_7''' $ Engl.) Vaginar. crassitudo, 0,0081—0,0112mm. $(_{\overline{2}\overline{8}_5}—_1^1{}_8''' $ Engl.) Diam. cellular. 0,0196— 0,0224mm. $(_1^1{}_0—_9^1{}_4$ Engl.) Trichomatum longitudo 8—15mm.

Inter alias algas natantes et affixas in dispersis trichomatibus. (DISTRIB. *L. majoris* in Europa orientali.)

Hujus formæ cellulæ cellulis interstitialibus non interruptæ cylindrum continuum formant, diametro trichomatum apicem versus non decrescente, ultima cellula late rotundata, vaginæ in trichomatum summis utplurimum sunt apertæ et cellulis vacuæ. In fere omnibus trichomatibus a me visis *Microthamnii* novi elegantis plantulas observavi, quæ *Lyngbyæ* sunt affixæ radiculis contortis sæpe circum circa trichoma procurrentibus.

Speciminum formæ typicæ ex mari Adriatico trichomata paulo sunt crassiora, sæpissime occurrunt cellulæ interstitiales colore rubro-lutescente distinctæ ceteris cellulis trichomatis.

1. **Limnactis minutula,** *Kütz.,* *Tab. Phyc. ii. tab.* 63, *fig.* 1 ; var. trichomatibus rectis sensim attenuatis margine crenulatis, cellulis distincte separatis diametro triplo-quadruplo brevioribus, cytioplasmate dense grossius granuloso, cellulis summis diametro usque quadruplo longioribus, hyalinis distinctis, vaginis hyalinis, cellulis perdurantibus sphæricis cellularum diametro æqualibus. Diam. trichomatum, 0,0056—0,0075mm. $(_5^1{}_{\overline{6}}—_2^1{}_{\overline{0}\overline{0}}''' $ Engl.)

In *Schizosiphontis kerguelensis* trichomatibus in cæspitulis parvulis usque 0,28mm. $(_4^1''' $ Engl.) latis. DISTRIB. Gallia, Germania, Dania, Succia, Britannia.

1. **Dasyactis Kunzeana,** *Kütz.* Diam. trichomat. 0,0056—0,0068mm. $(_{3\overline{7}\overline{8}}^1—_{3\overline{2}\overline{0}}^1''' $ Engl.) Diam. cum vagin. 0,0112mm. $(_{\overline{1}\overline{8}\overline{9}}^1''' $ Engl.)

In physcumatibus parvulis singulis dispersis in *Nitellæ antarcticæ* cellulis nidulantibus. (DISTRIB. Germania.)

1. **Mastigothrix articulata,** *Reinsch in Journ. Linn. Soc.* xv. 207 ; trichomatibus prolongatis subcylindricis basin versus paulo incrassatis distincte articulatis, articulis inferioribus indistinctioribus, superioribus loculamentis distincte disjunctis diametro subæqualibus, cytioplasmate granulis majoribus instructo, sporis perdurantibus obovalibus dimidio (et paulo minus) trichomatis latitudinis æquantibus. Diam. trichom. (in basi) 0,0168mm. $(_{\overline{1}\overline{2}\overline{6}}^1''' $ Engl.) Diam. trichom. (in superiore parte) 0,0112mm. $(_{\overline{1}\overline{8}\overline{9}}^1''' $ Engl.)

In singulis trichomatibus partim in superficie partim in strato summo physcumatis *Hormosiphontis leptosiphontis,* s. n., nidulantibus observatum.

Mastigothrichi fusco, Kützing, simillima in forma ac crassitudine trichematum distinguitur : cellulis distincte articulatis cellulis perdurantibus minoribus a cellula infima sejunctis. Cellularum *M. fusci* cytioplasma subtiliter granulosum, cellula perdurans diametro cellularum æquante, basi lata (interdum intus excavata) cellulæ infimæ trichomatis arctissime adpressa, cytioplasmate homogeneo.

2. **Mastigothrix æruginea,** *Kütz.* Trichomata vix discernenda a speci-

minibus Europæis in thallo *Chætophorarum* et *Nostochidis* nidulantibus. In singulis trichomatibus inter *Tolypothrichis flaccidæ*, Kütz., cæspitulos nidulantibus. In singulis trichomatibus cellulæ inferiores breviores et indistinctius disjunctæ. (Dis-TRIB. Germania.)

3. **Mastigothrix minuta**, *Reinsch in Journ. Linn. Soc.* xv. 207 ; trichomatibus distincte articulatis apicibus rectis, articulis inferioribus dimidio latitudine brevioribus (et paulo magis), sporis perdurantibus obovalibus usque subsphæricis diametro dimidio trichomatis latitudinis æquante. Latit. trichomatum (in basi) 0,0084—0,0097mm. $(_{2\frac{1}{8}3}—_{2\frac{1}{2}6}''')$ Engl.) Diam. sporæ perdur., 0,0041mm. $(_{5\frac{1}{6}7}''')$ Engl.)

Inter algas minores (*Leptothrix*, *Coleochæte*) in foliis muscorum aquatilium insidentes.

A *Mastig. æruginea*, Kütz., dimensionibus duplo magis minoribus distincta species. In trichomatibus singulis vaginæ infima pars paulo incrassata et lamellosa, sed cellula perdurans non inclusa a lamellis.

1. **Hydrocoleum Eatoni**, *Reinsch in Journ. Linn. Soc.* xv. 207 ; fasciculis liberis inter alias algas dispersis usque ad 18mms. longis in summis sensim attenuatis, trichomatibus olivaceo-viridibus (a latere visis), 8is—12is consociatis et leviter contortis subtilissime distincte articulatis, cellulis distinctis omnibus homogeneis, diametro quintuplo brevioribus, cytioplasmate dense punctulato, vaginis achrois membranaceis duris subtiliter lamellosis, trichomatum fasciculi latitudinis dimidio crassis. Diam. fasciculi (in medio parte) 0,056—0,086mm. $(_{3\frac{1}{8}}—_{2\frac{1}{3}}''')$ Engl.) ; (in summis) 0,0224—0,0333mm. $(_{9\frac{1}{4}}—_{6\frac{1}{3}}''')$ Engl.) Diam. trichomatum 0,0041—0,0056mm. $(_{5\frac{1}{6}7}—_{3\frac{1}{4}8}''')$ Engl.) Vagin. crassitudo 0,0028mm. $(_{7\frac{1}{5}7}''')$ Engl.)

Inter muscos aquaticos et aliis *algis* (*Vaucheria*, *Schizosiphon*) inmixtum.

Hoc *Hydrocoleum* consentit cum *H. helvetico*, Nägeli, in fasciculorum dispositione, sed differt dimensionibus fasciculorum quintuplo magis majoribus, trichomatum diverso colore et cellulis brevioribus.

Tab. IV. Fig. i.—1, fasciculi media pars $(\frac{3\,6\,2}{1})$;—2, fasciculi summa pars $(\frac{3\,6\,2}{1})$.

Nostoc hydrocoleoides, *Reinsch in Journ. Linn. Soc.* xv. 208 ; subtilissimum, physcumate iu modo *Hydrocoleorum* teretiformi prolongato peridermate distincto hyalino ciucto ex trichomatibus et rectis et paulo contortis (5is—10is) fasciculatim conjunctis formato, trichomatibus pallide ærugineis parallelis leviter contortis vaginulis hyalinis velatis, cellulis vegetativis post divisionem diametro paulo longioribus, cytioplasmate punctulato, cellulis perdurantibus ceteris paulo majoribus sphæricis in trichomatibus sparsis. Diam. trichomat. 0,0022—0,0028mm. $(_{0\frac{1}{4}0}—_{1\frac{1}{5}7}''')$ Engl.) Diam. fasciculi (in media parte) 0,0112—0,0224mm. $(_{1\frac{1}{8}0}—_{0\frac{1}{4}}''')$ Engl.)

Inter *Tolypothrix Nægelii*, Kütz., et in massa parvula *Diatomacearum* foliis muscorum aquatilium et *Nitellæ* adhærente.

Hæc plantula paradoxa secundum structuram et physcumatis et trichomatum

Nostochidis generis bonam speciem se ostendit. Trichomata integumento communi distinctissime clausa vix sunt discernenda a trichomatibus specierum singularum. Species unica *Nostochidis* generis hucusque cognita physcumate filamentosa, a ceteris speciebus physcumate sive plano sive sphærico sat distincta.

Tab. IV. Fig. iv.—1, physcumatis pars media $(\frac{7.2^{0}}{1})$;—2, physcumatis summa pars, trichoma singulum usque in apicem excurrens $(\frac{7.2^{0}}{1})$.

2. Nostoc polysaccum, *Reinsch in Journ. Linn. Soc.* xv. 208 ; physcumate coriaceo irregulariter sphærico et subreniformi colore subaureo-fusco magnitudine seminis sinapeos ad Pisi sativi, intus loculamentoso ac dissepimentis coloratis lamellosis et radialiter et transversaliter positis percurso, peridermate firmo coriaceo fuscescente, trichomatibus centralibus paulo flexuosis, cellulis sphæricis colore pallide olivaceo, cytiodermate distincte dupliciter striato, cellulis perdurantibus sphæricis ceteris cellulis paulo latioribus. Diam. cellular. 0,0011mm. $(\frac{1}{50};'''$ Engl.) Diam. cellular. perdurant. 0,0056mm. $(\frac{1}{31};'''$ Engl.) Diam. physcumatis, 2,5— 3mm.

Forma (an status peculiaris evolutionis ?). Physcumate ex trichomatibus brevioribus vaginis amplis hyalinis homogenis (in modo *Hormosiphontis*) inclusis laxissime cohærentibus exstituto.

Tab. IV. Fig. i.—1, Physcumatis sectionis transversalis pars usque ad peripheriam physcumatis se pertinens, vesiculæ trichomata includentes, radialiter dispositæ, parietes vesicularum subcoloratæ $(\frac{6.1}{1})$. 2, Formæ physcumatis pars peripheriæ, sectio transversalis ; physcuma ex vesiculis numerosissimis, trichomatibus singulis inclusis formatum, trichomata breviora in modo *Hormosiphontis* indumento crasso subhyalino inclusa $(\frac{6.1}{1})$. 3, Specimina plantulæ (in spiritu vini asservatæ) magnitudine naturali.

3. Nostoc polysporum, *Reinsch. in Journ. Linn. Soc.* xv. 208 ; physcumate sphærico magnitudine pisi minoris, indumento crasso hyalino distinctissime plurilamelloso velato, trichomatibus laxius positis subcontortis pallide ærugineis, cellulis sphæricis arctissime conjunctis, post divisionem transverse ellipticis, cellulis perdurantibus numerosissimis sphæricis ceteris cellulis duplo majoribus cytiodermate crasso. Diam. cellular. 0,0028mm. $(\frac{1}{717};'''$ Engl.) Diam. cellular. perdurantium 0,0041—0,0058mm. $(\frac{1}{501}-\frac{1}{310};'''$ Engl.). Diam. physcumatis, 3—4 mm.

Inter alias algas fluitans (in paucis speciminibus observatum).

A persimilibus : *N. gymnosphæricum* et *N. cœruleum*, Kützing, Tab. Phycol. ii. tab. 3, fig. 3, 4, differt indumento plurilamelloso, cellulis perdurantibus numerosioribus.

4. Nostoc species, e minoribus, physcumate irregulariter polyedrico, textura cartilaginea, colore rubro-fusco, trichomatibus contortis, cellulis subsphæricis arctissime adjacentibus, cellulis perdurantibus sphæricis ceteris cellulis paululo majoribus cytiodermate crasso distincto. Diam. cellular. 0,003—0,0011mm. $(\frac{1}{717}-\frac{1}{567};'''$ Engl.) Diam. physcumatis 1,8mm.

** ⁣ I

Inter *Zygnemam*. In a freshwater pool, Swain's Bay.

In textura, forma irregulari physcumatis minus in trichomatum forma *N. eduli* Berkeley persimile.

Specimen unicum observatum speciem accuratius constituendam mihi non permittit.

5. **Nostoc paludosum,** *Kütz.*, *Tab. Phyc. ii. tab.* 1, *fig. 2.* Specimina singula observata insidentia plantulis *Bulbochœtcis* foliis muscorum insidentibus et in trichomatum crassitudine et in cellularum forma maxime consentiunt cum speciminibus Germanicis et cum icone Kützingiana. Diam. cellular. trichomat. 0,0011—0,0018mm. $_{1}$$\frac{1}{800}$—$_{1}\frac{1}{300}$''' Engl.)

Ab omnibus *Nostochidis* speciebus pagnitis species cellulis minimis. DISTRIB. In Europa vulgaris.

6. **Nostoc leptonema,** *Reinsch in Journ. Linn. Soc.* xv. 209 ; physcumatibus usque semini sinapeos æqualibus sphæricis paulo elasticis arctissime conjunctis cohærentibus, indumento exteriore suberasso hyalino homogeneo, trichomatibus prolongatis multipliciter contortis laxius (in majoribus) et densius (in minoribus) intricatis, cellulis oblongis polis attenuatis laxe se adtingentibus ; cellulis perdurantibus sphæricis usque subovalibus sparsis ceteris cellulis duplo paulo magis majoribus. Diam. cellular. 0,0015—0,0024mm. ($\frac{1}{930}$—$\frac{1}{810}$''' Engl.) Diam. cellular. perdur. 0,0056mm. ($\frac{1}{318}$''' Engl.) Diam. physcumatis 0,2—1,5mm.

In muscorum caulibus et foliis physcumatibus cohærentibus, partim corpora urœformia formans.

A *Nostochidibus* physcumate sphærico *Nostoc aureum,* Kütz., Tab. Phyc. ii. tab. 1, fig. 4 (planta marina) proximum in magnitudine et textura physcumatis ac crassitudine trichomatum ; hoc *Nostoc* differt trichomatibus brevissimis valde curvatis cellulis perdurantibus minoribus.

Inveniuntur interdum muscorum foliis insidentia corpora ex parenchymatice conjunctis physcumatibus varia magnitudine formata.

Forma : *Crystallophorum.* Physcumate corporibus crystallisatis subsphæricis inclusis ex crystallis (Ferri oxydati ?) radialiter dispositis formatis. Diam. corpor. crystallisat. 0,0224—0,05mm. ($\frac{1}{91}$—$\frac{1}{42}$''' Engl.)

1. **Anabaina confervoides,** *Reinsch in Journ. Linn. Soc.* xv. 209; e subtilioribus stratum tenue formans, trichomatibus prolongatis rectissimis parallelis in mueo communi nidulantibus, cellulis distinctissimis rectangularibus usque subquadraticis, spatiis interloculatis angustioribus distinctis sejunctis, diametro transversali paulo longioribus (usque duplo), cytioplasmate subtiliter granuloso colore pallide ærugineo ; cellulis perdurantibus ellipticis ceteris cellulis paulo latioribus et longioribus. Diam. cellular. 0,0022—0,0028mm. ($\frac{1}{940}$—$\frac{1}{760}$''' Engl.)

In stratis tenuioribus inter alias Algas.

Hæc species peculiaris distinguitur ab omnibus hucusque cognitis speciebus cellulis angulosis (nec sphæricis nec ellipticis).

2. **Anabaina involuta,** *Reinsch in Journ. Linn. Soc.* xv. 299; libere natans, e tenuioribus, trichomatibus prolongatis multipliciter involutis, cellulis sphæricis et ellipticis (in statu indiviso) intermixtis, cytioplasmate subtiliter granuloso, cytiodermate extrorsum mueo hyalino tenui velato, cellulis perdurantibus (sporis) sphæricis sparsis ceteris cellulis paululo latioribus, cytiodermate crasso distincto, cytioplasmate granuloso. Diam. cellular. 0,0024—0,0032mm. $(_{9}\frac{1}{4}_{5}$— $_{7}\frac{1}{1}_{6}'''$ Engl.) Diam. cellular. perdurant. 0,0041mm. $(_{5}\frac{1}{6}_{1}'''$ Engl.)

In trichomatibus singulis inter alias algas Phycochromaceas interjectis.

Ab *A. circinali* in trichomatum forma persimili differt cellulis quadruplo magis minoribus, cytioplasmate subtiliter granuloso.

1. **Hormosiphon leptosiphon,** *Reinsch in Journ. Linn. Soc.* xv. 210; globosum, magnitudine Pisi sativi, olivaceo-viride, physeumate subcartilagineo intus molli, indumento exteriore subtenuee, trichomatibus prolongatis marginem physeumatis versus radialiter dispositis subcontortis pallide æerugin, indumento hyalino homogeneo decolorato subtenui velatis, cellulis subsphæricis subarete conjunctis, cellulis perdurantibus sphæricis ceteris cellulis duplo latioribus. Diam. cellular. 0,0028mm. $(_{7}\frac{1}{3}_{1}'''$ Engl.) Diam. cellular. perdurant. 0,0056mm. $(_{3}\frac{1}{4}_{8}'''$ Engl.) Diam. physeum. 3,5—6mm.

In physeumatibus singulis inter *Schizosiphontis kerguelensis* cespites.

Physeumatum observatorum dua procreant intus physeumata singula filialia ellipsoidea trichomatibus brevissimis subrectis et leviter contortis indumento crasso hyalino decolorato velatis densissime repleta, indumento communi distincto velata. Physeumatis externa pars plerumque ex trichomatibus physeumatum filialium trichomatibus simillimis formata. Superficies physeumatum est vestita plantulis variis egregie *Stigeoclonio subtili*, n. sp., singulis filis *Euactidis Künzeanæ* et *Tolypothrichis flaccidæ*.

Tab. IV. Fig. vii.—1, physeumata dua magnitudine naturali.—2, trichomatis singuli pars maxime aucta, α. cellula perdurans $(^2\frac{1}{1}^9)$.—3, physeumatis sectionis transversalis pars exterior, cum physeumate filiali singulo trichomatibus brevissimis indumento crasso velatis dense repleto, α. indumentum exterius physeumatis.

2. **Hormosiphon coriaceus,** *Kütz., Tab. Phyc.* ii., *tab.* 14, *fig.* 1; var. KERGUELENSIS, *Reinsch in Journ. Linn. Soc.* xv. 211; physeumate irregulariter expanso subplano subcoriaceo colore obscure rubro-fusco, in sectione transversali ex stratis 5is—7is formato, peridermate (sectionis transversalis physeum.) lateris superioris crassiore laniellose fusco, lateris inferioris peridermate tenuiore, trichomatibus vermiculiformibus multipliciter contortis, vaginis fuscis amplis distinctissimis pluri-lamellosis, cellulis sphæricis colore pallide æerugineo. Diam. cellular. 0,0041 mm $(_{7}\frac{1}{1}_{1}'''$ Engl.) Diam. trichom. (cum vaginis) 0,0224—0,0278 mm. $(_{9}\frac{1}{4}—_{7}\frac{1}{5}'''$ Engl.) Physeumatis crassitudo 0,139—0,168 mm. $(_{7}\frac{1}{4}—_{9}\frac{1}{2}'''$ Engl.)

Specimina majora in spiritu vini asservata ac in charta intenso.

Marshy ground near Vulcan Cove. (DISTRIB. Gallia, Germania, Italia.)

1. **Schizosiphon· kerguelensis,** *Reinsch in Journ. Linn. Soc.* xv. 211 ;
cæspitosus, cæspitulos confertos radialiter dispositos usque 6 mms altos in muscis
aquaticis affixos formans, trichomatibus radiantibus e basi repetito dichotomo-
ramosissimis in summis fastigiatis, pseudoramulis ultimis corymbosis fasciculatis
apicibus paulatim subangustatis, vaginis pseudoramulorum ultimorum fuscis integerrimis cellularum diametro subæqualiter crassis, vaginis trichomatum inferioris
partis crassioribus dense subtiliter lamellosis, cellulis omnibus æqualibus distinctis,
diametro subæqualibus apicem pseudoramulorum versus non decrescentibus, cytioplasmate colore pallide olivaceo-viridi granulis majoribus distinctis dense repleto,
cellulis perdurantibus singulis aut compluribus basilaribus subsphæricis diametro
cellularum æqualibus. Diam. trichomat. (in diversis locis mensuratus) 0,0108—
0,0333 mm. ($\frac{1}{1\,1\,5}$—$\frac{1}{3\,3}$″ Engl.)   Diam. pseudoramulorum ultimorum 0,013—0,0168
mm. ($\frac{1}{1\,3\,7}$—$\frac{1}{1\,2\,6}$″ Engl.)

Hab. in muscis aquaticis caules densissime pelicuhe formiter inducens.

Hæc species elegantissima in cæspitulis muscis in caliculo vitreo inclusis co-
piosissima est reperta, in primis speciminis majoris caules densissime erant obtecti.
Cellularum funiculus singulis locis haud raro et simpliciter et dupliciter contortus,
quæ partes trichomatum paulo sunt incrassatæ ; basin trichomatum versus cellularum funiculi plerumque sunt contorti ; trichomatum infimæ partis vaginæ pluri-
lamellosæ et trichomatum infima pars cuneiformiter angustata in filum singulum
producta.

1. **Tolypothrix flaccida,** *Kütz. Tab. Phyc.* ii. *tab.* 32, *fig.* 2.   Forma
cellulis diametro transversali æqualibus et paulo longioribus.   Diam. trichom.
0,0050—0,0084 mm. ($\frac{1}{3\,1\,8}$—$\frac{1}{2\,4\,3}$‴ Engl.)

In cæspitulis parvulis in foliis muscorum aquaticorum insidens.   (DISTRIB. *T.
flaccidæ*, Britannia, Gallia, Germania, Helvetia.)

Hæc formæ sunt peculiares ut in forma typica, cellulæ perdurantes complures
postpositæ, cellulæ complures funiculi trichomatum sæpissime interstitiis hyalinis
sunt disjunctæ et trichomatum summa pars cellulis vacua.

2. **Tolypothrix Nægelii,** *Kütz.*

Hæc *Tolypothrix* a forma typica est distincta trichomatibus paulo tenuioribus,
pseudoramulis crebrioribus, quæ sunt brevissimæ in singulis trichomatibus ; summa
pars cellulæ perdurantis singulæ in pseudoramulorum basi nonnunquam est trun-
cata.

Inter *Schizosiphontis* cæspites et affixi et fluctuantes cæspituli.   (DISTRIB. *T.
Nægelii*, Helvetia.)

1. **Schizothrix hyalina,** *Kütz. Spec. Alg. Tab. Phyc.* ii. *tab.* 40, *fig.* 1.
Var. RAMOSISSIMA, *Reinsch in Journ. Linn. Soc.* xv. 211 ; trichomatibus *Schizo-
siphonti* insidentibus subtilissimis, funiculis et submoniliformibus et subcylindraceis
pallide ærugineis, vaginulis amplis hyalinis cinctis ; pseudoramulis numerosis erectis
flagelliforme attenuatis. Diam. trichom. (cum vaginulis) 0,0022—0,0056 mm.

($_{a}\frac{1}{u}$—$_{a}\frac{1}{7}$''' Engl.) Diam. trichom. intern. 0,0011 mm. ($_{1}\frac{1}{9}$''' Engl.) Altitudo plantulæ 0,8 mm. ($\frac{1}{4}$''' Engl.)

In *Schizosiphontis kerguelensis* trichomatibus in cæspitulis dispersis. (DISTRIB. *S. hyalinæ*, Montibus Europæ.)

Hanc formam peculiarem, verisimile speciem propriam, tantummodo in paucis sed bonis speciminibus observavi, quæ erant apta ad constituendum genus. Est similitudo maxima cum *Schizothr. hyalina* in trichomatum et vaginarum crassitudine et cellularum funiculi forma, quamquam incrementi modus et loci natalis est diversissimus.

1. **Sirosiphon vermicularis**, *Reinsch in Journ. Linn. Soc.* xv. 211; e minimis, cæspitulis parvulis trichomatibus subrectis summis attenuatis procumbentibus intertextis, plus minusve ramosis, ramulis alternantibus apicem versus sensim attenuatis ramulis summis diametri trichomatis primarii dimidio tenuioribus, trichomatum cellulis uniseriatis arctissime conjunctis, cytiodermate subcrasso firmo fuscescente, cytioplasmate subtilissimo granuloso, ramulorum cellulis apicem ramulorum versus angulosis confervoideis, vaginis (trichomat. primarior.) tenuioribus (vix cellular. diametri octavam partem) simpliciter striatis cellulas arctissime includentibus; cellulis interstitialibus nullis. Diam. trichom. primar. 0,0112 mm. ($_{1}\frac{1}{89}$''' Engl.) Diam. ramulorum 0,0056 mm. ($_{a}\frac{1}{78}$ Engl.)

In cæspitulis singulis inter alios *Sirosiphontes Hormosiphonti coriaceo* prope Vulcan Cove adhærentes.

Ab omnibus *Sirosiphontibus* hucusque cognitis species minutissima. *Sirosiphon* in ramulorum cellulis diversis a cellulis trichomatis primarii. *Sirosiphonti sylvestri*, Itzigsohn. proximus sed sat distinctus trichomatibus tenuioribus cellulis cytiodermate tenuiore indistincte articulatis.

2. **Sirosiphon pulvinatus**, *Kütz.;* var. cellulis trichomatis primarii cytiodermate crassissimo colorato absque ordine biseriatis cellularum ramulorum uniseriatis aut absque ordine biseriatis. Diam. cellular. 0,0056—0,0068 mm. ($_{5}\frac{1}{4}$—$_{3}\frac{1}{86}$ Engl.) Diam. cellular. c. vagina 0,013 mm. ($_{1}\frac{1}{57}$''' Engl.) Trichomat. crassit. 0,0224—0,0306 mm. ($_{9}\frac{1}{4}$—$_{9}\frac{1}{9}$''' Engl.) Trichomata perpauca dispersa. Forsan propria species. (DISTRIB. *S. pulcinati*, Europa, Americ. boreal.)

Var.; trichomatibus irregulariter ramosis, ramulis apice obtusis numerosis subcontortis, cellulis omnibus æqualibus subovalibus, cytiodermate tenuiore hyalino decolorato, irregulariter biseriatis. Dimensionibus iisdem præced.

3. **Sirosiphon** species nova, *Reinsch in Journ. Linn. Soc.* xv. 212; e minoribus, trichomatibus singulis inter alias algas dispersis, irregulariter pinnatoramosis, ramis bilateralibus trichomati primario æqualiter formatis et æqualiter crassis, summis non attenuatis, cellulis subsphæricis spatiis hyalinis disjunctis, cytiodermate tenui homogeneo subhyalino, cytioplasmate subhomogeneo pallideæruginoso, vaginis crassis hyalinis subhomogeneis decoloratis, cellulis interstiti-

alibus? Diam. cellular. 0,0041—0,0056 mm. ($\frac{1}{5\,0\,0}$—$\frac{1}{3\,1\,8}$''' Engl.)   Trichomat. crassit. 0,0112—0,013 mm. ($\frac{1}{1\,8\,0}$—$\frac{1}{1\,3\,7}$''' Engl.)

In trichomatibus singulis inter alias *Sirosiphontes* et inter *Scytonemam castaneum* inter *Hormosiphontem coriaceum* (near Vulcan Cove). *S. celulinus* et *S. hormoides* Kützing trichomatibus crassioribus fasciculatoramosis et dichotome ramosis distincti. *S. panniformis*, Kütz., distinguitur ramis elongatis trichomate primario tenuioribus et cellulis interstitialibus.

4. **Sirosiphon kerguelensis**, *Reinsch. in Journ. Linn. Soc.* xv. 212 ; trichomatibus ramosissimis, trichomate primario procumbente ramis irregulariter ramosis ramulis ultimis apicem versus æqualiter latis, cellulis trichomatis primarii ac ramulorum ovalibus usque irregulariter sphæricis in seriem simplicem dispositis, intervallis hyalinis usque cellularum longitudini æquantibus disjunctis, articulis tubuliformibus angustissimis (lacunis tubuliformibus in muco vaginæ) conjunctis, cytioplasmate subhomogeneo dilute ærugineo, cytiodermate subtili decolorato (cellularum trichom. primarii crassiore fuscescente), cellulis summis ramulorum cohærentibus lyngbyaecis, vaginis crassis hyalinis decoloratis subhomogeneis (vaginis trichom. primarii sublamellosis anreis). Diam. trichom. primar. 0,0278—0,0333mm. ($\frac{1}{7\,5}$,—$\frac{1}{6\,0}$ Engl.) Diam. ramulorum 0,0248 mm. ($\frac{1}{8\,1}$''' Engl.) Diam. cellular. 0,013 mm. ($\frac{1}{1\,3\,7}$''' Engl.)

In trichomatibus singulis inter alias *Sirosiphontes*. Cum præcedente.

Hic *Sirosiphon* primo pro formam propriam *Sirosiphontis ocellati* habitus, eni est persimilis in trichomatis ramificatione et crassitudine, sed propter propriam de ceteris *Sirosiphontibus* discedentem structuram trichomatis propriam speciem se offert.

Tab. IV. Fig. vi.—1, trichomatis pars summa $\frac{4\,6\,0}{1}$;—2, trichomatis pars maxime aucta, $\frac{2\,6\,0}{1}$.

In singulis speciminibus observavi *Sirosiphontem* sequentem quem hujus *Sirosiphontis* varietatem puto. Trichomata ramosa ramis subintegris adscendentibus, cellulis ovalibus usque subsphæricis, intervallis hyalinis disjunctis. Articuli tubuliformes cellulas singulas conjungentes plurimum desunt.

5. **Sirosiphon Oliveri**, *Reinsch in Journ. Linn. Soc.* xv. 213 ; cæspitulis parvulis, trichomatibus adscendentibus prolongatis subramosis, ramulis singulis (et raro ramulis complnribus brevioribus approximatis) et leviter contortis, e serie simplice cellularum formatis, cellulis ovalibus diametro dimidio brevioribus (et paulo magis et minus), cytiodermate firmo crasso extrorsum fuscescente, cytioplasmate subhomogeneo obscure-ærugineo, vagina membranacea simplici subtenui, cellulis interstitialibus nullis. Diam. trichomatum (cum vaginis) 0,0196—0,0221 mm. ($\frac{1}{5\,1\,0}$—$\frac{1}{5\,4}$''' Engl.)

In cæspitulis parvulis inter *Hormosiphontem coriaceum* cum cæspitulis *Scytonematis castanei* intermixtis ; cum præced.

S. celutino, Kütz., et *S. hormoide*, Kütz., speciebus proximis in cellularum forma ac dispositione differt trichomatibus subintegris, vaginis tenuioribus.

Tab. IV. Fig. ii.—1, trichomatis summa pars ($\frac{140}{1}$);—2, trichomatis pars maxime aucta, vagina dupliciter striata, cellularum cytioderma dupliciter striatum, cellula singula longitudinaliter divisa ($\frac{730}{1}$).

6. **Sirosiphon secundatus,** *Kützing, Tab. Phycol.* ii. *tab.* 37, *fig.* 1; forma trichomate primario partim incrassato, ramis prolongatis apice incrassato; cellulis parvulis trichomatis primarii numerosis absque ordine dispositis, cellulis ramorum uni- aut irregulariter biseriatis cytiodermatibus crassis confluentibus. Diam. trichomatis primarii 0,0333—0,0393 mm. ($\frac{1}{30}$—$\frac{1}{34}$''' Engl.) Diam. ramorum 0,0221—0,0278 mm. ($\frac{1}{51}$—$\frac{1}{40}$''' Engl.) Diam. cellular. 0,0056 mm. ($\frac{1}{312}$''' Engl.)

In specimine singulo observato, inter alias *Sirosiphontes*. Cum præcedente. (DISTRIB. Europa.)

CHLOROPHYLLOPHYCEÆ.

1. **Glœocystis vesiculosa,** *Nægeli.* Cellulæ indivisæ usque ad 0,011 mm. ($\frac{1}{160}$''' Engl.) diam.; familiæ bicellulares 0,0058 mm. ($\frac{1}{318}$''' Engl.) diam.

Inter algas unicellulares adhærentes foliis muscorum. (DISTRIB. Germania, Helvetia.)

1. **Palmella mucosa,** *Kütz. Tab. Phyc.* i., tab. 16, fig. 7; cellul. diam. 0,0056—0,0112 mm. ($\frac{1}{318}$—$\frac{1}{160}$''' Engl.)

Inter alias algas unicellulares. (DISTRIB. Europa.)

Distinguitur a forma communi cellulis paulo minoribus et integumentis crassioribus distinctius limitatis.

1. **Pleurococcus vestitus,** *Reinsch, Algenfl. Frank.*, p. 56, tab. iii., fig. 1). Var. MINOR; cellulis sphæricis singulis aut binis aut quaternis et compluribus sphærice conjunctas familias formantibus, cytioplasmate dense subtiliter granuloso, cytiodermate crasso (interdum colorato) verruculis acutis dispersis instructo. Diam. cellular. 0,0112—0,013 mm. ($\frac{1}{180}$—$\frac{1}{137}$''' Engl.)

Inter alias algas unicellulares. (DISTRIB. *P. vestiti*, Germania).

2. **Pleurococcus angulosus,** *Corda.* Protoc. palustris *Kütz. Tab. Phyc.* i. *tab.* 9; forma, cellulis sphæricis in familias minores in modo *Merenchymatis* cohærentes collocatis. Cellular. diam. 0,0041—0,0056 mm. ($\frac{1}{487}$—$\frac{1}{314}$''' Engl.) Diam. familiar. 0,0221—0,0278 mm. ($\frac{1}{34}$—$\frac{1}{36}$''' Engl.)

Cum præcedente. (DISTRIB. *P. angulosi*, Europa.)

Scenedesmus acutus, *Meyen.*

In singulis speciminibus observatum inter *Zygnemæ* cæspites; in a freshwater pool on the W. of Swain's Bay. (DISTRIB. Europa.)

1. **Botryococcus Braunii,** *Kütz.*

Maxime consentit cum speciminibus Europæis e diversis locis. Inter cæspites *Schizosiphontis* et in massis parvulis algarum unicellularium muscis aquaticis adhærentibus. Inveniuntur familiæ et virides et fuscescentes. (DISTRIB. Germania, Helvetia.)

1. **Oocystis Nægelii**, *Al. Braun.* Longit. cellular. 0,0278—0,0306 mm. ($\frac{1}{75}$—$\frac{1}{80}$''' Engl.) Latit. cellular. 0,0168 mm. ($\frac{1}{134}$''' Engl.) Magnitudine ac forma cellularum ac cytioplasmatis textura maxime consentiunt specim. cum speciminibus Germanicis. Indumentum familiarum bi- aut quadricellularium distincte dupliciter striatum.

In singulis familiis in massa parvula *Phycochromophyccarum* unicellularium *Hormosiphonti* adhærente; marshy ground near Vulcan Cove. (DISTRIB. Germania).

1. **Dictyosphærium Ehrenbergii**, *Nægeli;* cellulis paulo majoribus. Diam. cellular. 0,0084—0,0112 mm. ($\frac{1}{270}$—$\frac{1}{189}$''' Engl.) Inter algas varias muscorum foliis insidentes. (DISTRIB. Europa meridionalis.)

1. **Pediastrum ellipticum**, *Ralfs Brit. Desmid.;* var. ÆQUILOBUM ; cœnobio elliptico continuo, cellulis disci regulariter 5-6-gonis, membrana hyalina achroa lævissima, cellulis periphericis leviter obtusangulo-emarginatis, lobulis æqualibus cellulæ dimidio brevioribus apice truncatulis. Longit. maxima cœnobii 0,278— 0,336 mm. ($\frac{1}{8}$—$\frac{1}{7}$''' Engl.) Diam. cellular. 0,0278—0,032 mm. ($\frac{1}{75}$—$\frac{1}{64}$''' Engl.) In speciminibus duobus inter *Hormosiphontis* physcumata observatum. (DISTRIB. *P. elliptici*, Britannia.)

Asterosphærium,* genus novum *Protococcacearum.* Cœnobium sphæricum, intus excavatum, libere natans, e cellulis angulosis parenchymatice arctissime conjunctis (sient in *Pediastris*), extrorsum pyriforme ampliatis et subito angustatis formatum.

1. **Asterosphærium elegans**, *Reinsch in Journ. Linn. Soc.* xv. 213. Cœnobium sphæricum e cellulis 64 ant 128 formatum. Diam. cœnobii ex cellulis 128 formati 0,144 mm. ($\frac{1}{35}$''' Engl.) Inter algas minores libere natans (in paucis speciminibus observatum.)

Hoc genus proxime se continuatur generibus *Protococcacearum* cœnobio ex cellulis parenchymatice conjunctis formato (*Hydrodiction, Pediastrum, Cœlastrum, Staurogenia*). Cœnobii dispositio fit in quoque hemisphæra secundum seriem: 1, 6, 11, 16, 21, 26 (seriem arithmeticam primæ ordinis cum numero differentiali=5). Quo dispositionis modo hoc genus discedit a *Pediastris*, generi proximo. *Pediastrorum* plurimum specierum dispositio cœnobii fit, in speciminibus regulariter formatis, secundum seriem : 1, 5, 10, 16 (seriem arithmeticam secundæ ordinis cum serie differentiali prima: 4, 5, 6, et numero differentiali=1).

In *Asterosphærii* cœnobiis legem dispositionis cellularum, ut fere fit, ad explicanda cœnobia pervenire in omnibus casibus, certissime puto; sicut per analogiam in *Protococcaccis* cœnobio pluricellulari sphærico (*Cœlastrum et Sorastrum*), quorum cœnobia abnormiter disposita rarissime observari possunt.

* ἀστήρ stella, σφαῖρα globus.

Omnium speciminum observatorum cellulæ erant vacuæ, veluti sæpe observamus in *Pediastris* majoribus.

Tab. IV., Fig. viii.—1, specimen integrum ex cellulis 128 exstitutum ($\frac{128}{1}$);— 2, cœnobii marginis pars magis aucto ($\frac{128}{1}$).

1. **Glœocystis botryoides**, *Nægeli, Gatt. einzell. Alg.* Cellular. diam. 0,0022—0,001 mm. ($\frac{1}{910}$—$\frac{1}{567}$''' Engl.) Thallus gelatinosus, cellulis singulis et quaternatis, tegumentis crassis hyalinis distinctis.

In massis parvulis cum aliis algis *Phyeochromaceis* inter *Hormosiphontem coriaceum* var., prope Vulcan Cove. (DISTRIB. Europa orientalis.)

2. **Glœococcus** species. Diam. cellular. 0,0041—0,0056 mm. ($\frac{1}{345}$; — $\frac{1}{345}$''' Engl.) Cellulæ subsphæricæ in familiis 4- et 8-cellularibus consociatæ, cytioplasmate colore intensive viridi, locello hyalino decolorato singulo instructo.

In familiis singulis dispersis inter alias algas *Hormosiphonti coriaceo*, var. adhærentes.

1. **Polyedrium tetrætricum**, *Nægeli.* Cellulæ angulis acutiusculis (vix aculeolatis), marginibus lateralibus subrectis. Diam. cellular. 0,0224 mm. ($\frac{1}{64}$''' Engl.)

In speciminibus singulis inter alias algas unicellulares *Hormosiphonti coriaceo* var. adhærentes, prope Vulcan Cove. (DISTRIB. Europa australis.)

2. **Polyedrium minimi**, *Al. Braun, Alg. unicellut.*, p. 94, forma; cellulæ regulariter tetragonæ (quadraticæ) marginibus lateralibus omnibus æqualibus (vix leviter repandis), angulis obtuso rotundatis. Latit. 0,0006—0,0075 mm. ($\frac{1}{315}$—$\frac{1}{266}$''' Engl.)

In speciminibus singulis in massa parvula algarum *Zygogonio toruloso* var. adhærente. (DISTRIB. *P. minimi* Europa orientalis.)

Polyedrium Pynæidium, Reinsch, Algenflora von Franken. 1866, p. 80, tab. iii. a.—d., complures formas comprehendit. Specimen fig. *d.* delineatum est *P. minimum* Al. Braun, "lateribus alternis profundius emarginatis;" specimen fig. *a.* delineatum cum speciminibus Kerguelensibus exacte consentit; specimen fig. *b.* formam repræsentat marginibus æqualiter emarginatis.

1. **Chlamydococci** species, *Reinsch in Journ. Linn. Soc.* XV. 214; cellulis globosis vel ellipticis magnitudine paulo diversis, cytioplasmate et subhomogeneo et granuloso (granulis amylaceis dense repleto), in statu progressiore corpusculis sphæricis majoribus colore intensive luteo-purpurascente binis-quinternis instructo (cellulis filinlibus, Zoogonidiis), cytiodermate hyalino crassissimo plurilamelloso (interdum unilateraliter incrassato). Cellular. diam. (ante divis.) 0,0278—0,0393 mm. ($\frac{1}{715}$ — $\frac{1}{514}$''' Engl.) Diam. post divisionem 0,0196—0,0224 mm. ($\frac{1}{1010}$ — $\frac{1}{864}$''' Engl.)

Hab. in foliis musci aquatici.

Hujus plantulæ vera natura initio mihi erat aliquid dubia. In cellularum plurimum magnitudine cellulas filias non procreantium, cytioplasmatis colore, cytio-

**   K

dermatis structura valde consentiens cum *Chroococco aureo*, nihilominus inveniuntur cellulæ singulæ cytioplasmatis valde diversa structura a *Chroococcis*. Sed
post observatis cellulis minoribus (Zoogonidiis) sine dubio in cohærentia organica
cum cellulis majoribus *Chroococcoideis*, hujus plantulæ positio in systemate est
constituta. Quæ cellulæ forma late pyriformi, polo subito angustato, cytiodermate
tenui, cytioplasmate homogeneo colore intensive purpureo erant inventæ in con-
sortio cellularum majorum cytioplasmate vacuarum. In singulis cellulis sunt
inclusæ complures cellulæ filiales sphæricæ colore luteo-purpureo, aliis cellulis sunt
corpuscula bina (interdum singulum, cellula super.). Nonnullarum cellularum
cytioplasma densissime est repleta corpusculis amylaceis. *Chlamydococci* species
duæ cognitæ differunt cytiodermate multo tenuiore non lamelloso. Cellulæ filiales
duæ ($\frac{3\,4}{1}$°); cellula singula, cytiodermate unilateraliter incrassato, cytioplasmate
cellulis filialibus (*gonidiis*) α. compluribus ($\frac{4\,4\,4}{1}$); cellula singula cytiodermate
tenuiore, cytioplasmate corpusculis amylaceis densissime repleto ($^{3}\frac{4}{1}$°). Cellula
minor pyriformis (*Zoogonidium*).

1. **Cosmarium pseudo-nitidulum**, *Nordstedt, Bydr. till Kaenned. om
sydl. Norges Desmid. Lund.* 1872, tom. ix., p. 46, tab. i., fig. 4); var. semicellularum
semicircularium cytioderma in apice intus nodulo singulo incrassatum. Longit.
0,033mm. ($\frac{1}{0\,3}$''' Engl.) Lat. 0,0248 mm. ($\frac{1}{0\,2}$''' Engl.)

In speciminibus singulis in massis minoribus algarum variarum in muscis
aquaticis adhærentibus.

2. **Cosmarium crenatum**, *Bréb.* var. kerguelense; cellula in ambitu
late ovali, diametro longitudinali diametro transversali paulo longiore ($\frac{6}{5}$), semicellulis subsemicircularibus basi arctissime se adtingentibus incisura non disjunctis,
margine undato exciso, gibberulis truncatulis 14is—15is instructo, superficie verruculis in seriebus radialibus dispositis verruculosa, areola media lævi, semicellulis e
vertice visis ambitu ellipticis (in laterum medio leviter tumidis), isthmi latitudo
$\frac{1}{2}$ diametri transversalis. Diam. transv. 0,033 mm. ($\frac{1}{0\,3}$''' Engl.); diam. longit.
0,039 mm. ($\frac{1}{0\,4}$''' Engl.) Isthmi latitudo 0,0067 mm. ($\frac{1}{3\,4\,6}$''' Engl.)

In specimine singulo observatum inter *Foucheriæ* et *Schizothrichis* cæspites.
(DISTRIB. Europa, America borealis, Greenlandia.)

Formis singulis *Cosm. pulcherrimi* Nordstedt (Symb. ad Flor. Brasil. Desmid.
p. 175. tab. iii., fig. 24) simillimum in semicellularum ambitu et forma (*C. pul-
cherrim. β. boreale*, Nordst. Desmid. Spetsberg. et Beeren Eiland. p. 32, tab. vi.
fig. 14), sed differt semicellulis e vertice visis in medio utrinque non productis, a
fronte visis in medio lævibus.

1. **Staurastrum kerguelense**, *Reinsch in Journ. Linn. Soc.* xv. 214;
semicellulis a latere late trapezicis angulis longe productis, margine terminali sub-
recto a vertice visis regulariter trigonis, marginibus lateralibus rectis angulis in
cornulum rectum margine regulariter crenulatum longe productis, cornulis summis
bispinosis, cytiodermate lævi seriebus tribus verrucularum marginibus semicellulæ

parallclis et in cornulis excurrentibus ornato, isthmi latitudine quinta pars cellulæ latitudinis. Latit. cellulæ 0,1038 mm. ($\frac{1}{40}$''' Engl.); isthmi latitudo 0,0146 mm. ($\frac{1}{61}$''' Engl.)

Observavi tantum specimina dua in massa algarum muscis aquaticis adhærente.

S. gracili, Ralfs. simile semicellularum forma, sed differt dimensionibus duplo magis majoribus, cornulis multo gracilioribus.

E Familia pulcherrima *Desmidiaccarum* specierum numerosissima sunt reperta tantum *Cosmaria* dua hac *Staurastrum Palmoglæœ* species et *Euastrum binale* var.

1. **Euastrum binale**, *Turpin, var.* GIBBOSUM; semicellulis in sciagraphia trapezicis, margine terminali recto in medio levissime emarginato, angulis obtusis non productis, marginibus lateralibus gibberulis binis æqualibus rotundatis, superficie semicellulæ in quoque latere gibberulis binis æqualibus instructa, semicellulis a latere apice truncatis. Longit. 0,0306 mm. ($\frac{1}{85}$''' Engl.) Latit. 0,0224 mm. ($\frac{1}{94}$''' Engl.) Latit. margin. termin. 0,013 mm. ($\frac{1}{117}$''' Engl.) Isthmi latitudo 0,0041 mm. ($\frac{1}{587}$''' Engl.)

In singulis speciminibus inter algas unicellulares *Hormosiphonti* adhærentes. (DISTRIB. Europa, America borealis.)

A ceteris formis *Euastri binalis* hæc forma distinguitur superficie gibbosa semicellularum. *Euastrum binale var. dissimile*, Nordstedt (Desmid. Arctoæ, Konigl. Wetensk. Akad. Forhandl. Stockholm 1875, Nr. 6, p. 31, tab. viii. fig. 31), persimile in semicellularum sciagraphia, differt lobulis basalibus leviter repandis, angulis paulo productis, superficie non gibbosa.

1. **Palmoglœæ** species; cellulis ellipticis polis angustatis, diametro transversali dimidio diametri longitudinalis breviore, cytiodermate suberasso, cytioplasmate granulis singulis majoribus instructo in massa gelatinosa irregulariter expansa nidulantibus. Longit. cellular. 0,0068—0,0084 mm. ($\frac{1}{340}$—$\frac{1}{283}$''' Engl.) Latit. cellular. 0,0041 mm. ($\frac{1}{587}$''' Engl.)

In massis parvulis *Hormsiphonti* adhærentibus.

Granulis amylaceis cytioplasmatis ad *Palmoglœas* spectans, a *P. macrococca* et *micrococca* distinguitur cellulis minoribus et polis angustatis.

1. **Vaucheria sessilis**, *Vaucher.* Maxime consentit cum speciminibus Europæis. Oosporæ maturæ membrana trilamellosa.

Filum unicum fructiferum observari potuerat in massa ex algis diversis composita. (DISTRIB. Europa, America borealis.)

2. **Vaucheria sericea**, *Lyngbye.* Filum singulum florescens observatum. Oogonium ad fecundationem aptum, autheridia bina horizontaliter flexa nondum aperta. In filo singulo observato oosporas maturas evolvente oosporæ in oogonio laxe inclusæ. (DISTRIB. Europa, America borealis.)

3. **Vaucheria pachyderma**, *Synon.* VAUCH. DILLWYNI, *Web. et Mohr, exp.*

K 2

Fila eompluria oosporis maturis observata. Oosporarum membrana plurilamellosa duplo crassior membrana *V. sessilis*.

In cæspitulo parvulo ex filis intertextis *Vaucheriæ* specierum variarum composito inter *Nitellæ* specimina incluso, pauca observavi fila quæ pertinerent ad aliquam *Vaucheriam* ad Corniculatas spectantem. (DISTRIB. Europa frequens.)

De antheridiis et ogoniis nondum evolutis non potuerat discerni aliquid certi, nescioque, hæc fila pertinere ad *V. sericeam*, pachydermam an ad speciem propriam.

4. Vaucheria geminata, *De Candolle.*

Fila compluria oogoniis immaturis sine dubio ad *V. geminatam* spectantia; thalli ramulus lateralis minutus flores evolvens paulo longior et gracilior ramulo speciminum Europæarum, cornulum jam in positione propria, summo—ad observatorem verso—minime lateraliter contorto. *V. hamatæ* ramulus lateralis dimidio brevior ac ramuli oogonia procreantes duplo longiores. Antheridium a basi curvatum in uno anfractu contortum. (DISTRIB. Europa, America borealis.)

Status evolutionis partium florum *Vaucheriarum* perfecte congrunt cum eodem statu evolutionis florum *Nitellæ antarcticæ*. Tempus anni, respondens statu analogo vitæ harum plantularum in nostris latitudinibus ver est (menses Aprilis, Maii, usque ad initium mensis Junii). Quarum plantularum phænomena vitalia normam dare ad dijudicandas ullæ regionis terræ rationes in respectu commutationum temporum quadripartitarum anni, verisimile videtur.

1. Olpidium caudatum, *Reinsch in Journ. Linn. Soc.* xv. 215;

cellulis sphæricis sine radiculis substrato viventi insidentibus, in polo processu singulo spini formi cellulæ diametro subæquante postremo aperto instructis, cytiodermate distincto subcrasso, cytioplasmate dense granuloso.—Diam. cellular. 0,0112— 0,013 mm. ($\frac{1}{T h 9}$—$\frac{1}{1 5 T}$''' Engl.)

In *Schizosiphontis kerguelensis* trichomatibus.

O. ampullaceum (*Chytridium ampullaceum*, A. Braun, Ber. d. Berlin. Acad. 1855, p. 66; Rabenhorst, Fl. Eur. Alg. ii. p. 282) est distinctum ab hoc *Olpidio* dimensionibus duplo minoribus (0,0064 mm.; $\frac{1}{3 1 2}$''' Engl. diam.).

Tab. IV., Fig. vi.—1, *Schizosiphontis* trichomates pars cum plantula parasitica insidente, $\frac{350}{1}$;—2, cellula singula parasitica major amplificata, $\frac{720}{1}$.

1. Chytridium pyriforme, *Reinsch in Journ. Linn. Soc.* xv. 215;

cellulis zoogonidiis nondum egressis operculose apertis ovato-pyriformibus, basi sensim angustata, in radiculum in substrato vivente radicantem prolongatis diametro transversali dimidio diametri longitudinalis angustiore, cytioplasmate dense subtiliter granuloso, cytiodermate distincto dupliciter striato, cellulis zoogonidiis egressis subcylindricis usque subcuneatis, operculo transversaliter a cellula se sejungente subhemisphærico apice rotundato (non acuminato), radiculo usque tertiam partem diametri longitudinalis cellulæ æquante, in medio plus minusve incrassato apiculo prolongato deorsum verso. Diam. transvers. cellulæ 0,013 —

0,0168 mm. ($\frac{1}{134}$—$\frac{1}{58}$''' Engl.) Diam. longitud. cellulæ 0,0258—0,0278 mm.
($\frac{1}{16}$—$\frac{1}{15}$''' Engl.)

In *Vaucheriæ* cellulis.

A *Chytridiis* cognitis proximum in cellularum forma *Chytr. Olla*, A. Braun,
(Verjung. p. 198. Ber. Berlin. Acad. 1855, p. 380; Rabenh. Fl. Eur. ii. p. 277),
quod *Chytridium* distinguitur cellulis latioribus, operculo obtuse umbilicato;
C. acuminatum et *C. brevipes*, A. Braun, sunt distincta operculis acuminatis. In
omnibus *Vaucheriæ* cellulis, quæ portaverunt plantulas parasiticas, sunt observatæ
prolongationes utriculiformes dense positæ e *Vaucheriæ* cellula egressæ. Parasita
invenitur plerumque in iisdem locis *Vaucheriæ* cellulæ infectæ ubi sunt evolutæ
hæc prolongationes utriculiformes. In his locis abnormiter transmutatis apparent
parietes intercalares quæ separant lumen transmutatum a cellulæ cetero lumine.
Certissime adducta est transmutatio abnormis *Vaucheriæ* a plantulis parasiticis.
Complures casus hucusque sunt observati, in quibus efferuntur transmutationes
morphologicæ plantarum altiorum per plantulis parasiticis unicellularibus.* Singula
fila *Vaucheriæ* abnormiter transmutatæ observavi, quibus desunt *Chytridii* cellulæ,
sed in cytioplasmate *Vaucheriæ* sunt impositæ cellulæ sphæricæ magnitudine varia
manifesto alienæ *Vaucheriæ* cellulæ. Utrum aliquam connexionem esse geneticam
inter *Chytridii* cellulas pyriformes *Vaucheriæ* insidentes et cellulas entophyticas,
an non, incertum est.

1. **Microthamnion cladophoroides**, *Reinsch in Journ. Linn. Soc.* xv. 216;
e maximis, fruticulosum, filis solitariis erectis regulariter ramosis, radiculis singulis
contortis in substrato (algis viventibus) insidentibus, ramulis erecto-patentibus
attenuatis unilateraliter dispositis (in speciminibus minoribus) aut verticillatim
dispositis (in speciminibus majoribus), cellulis fili primarii apicem versus paulo incrassatis diametro 4plo-6plo longioribus, cellulis ramulorum in basi paulo con-
strictis diametro 10plo-20plo longioribus, cytioplasmate omnium cellularum subhomogeneo, colore pallide luteo-olivaceo, granulis singulis dispersis instructo. Fili
primarii cellularum latit. 0,0056 mm. ($\frac{1}{3}\frac{1}{8}$''' Engl.) Ramulorum cellularum latit.
0,0028—0,0041 mm. ($\frac{1}{7}\frac{1}{8}$—$\frac{1}{5}\frac{1}{8}$''' Engl.) Plantulæ altit. 0,556 mm. ($\frac{1}{4}$''' Engl.)
In *Lyngbyæ majoris*, Kütz. forma trichomatibus et in *Chlorococci* spec. cellulis
radiculis brevissimis affixum. Hæc plantula elegantissima *Cladophoris* singulis
in habitu haud dissimiles, sed sat distincta a *Cladophoris* cytioplasmate subhomogeneo ac dimensionibus minimis, ad *Microthamnia* spectat quibuscum consentit in
cytioplasmatis structura. Generis specierum trium hucusque cognitarum nulla
aliqua similitudine consentit cum plantula Insulæ Kerguelensis.

1. **Stigeoclonium Hookeri**, *Reinsch in Journ. Linn. Soc.* xv. 216;
læte viride, parasiticum, cæspitulos chætophoræformes basi radicante formans;
filis ætate provectiore inferne nudæ et subintegræ superne ramosissimis, basi

* *Synchytrium Taraxaci* (De Barg. et Woron. Ber. d. Naturf. Gesellsch. Freiburg, iii. 2. tab. i. ii.,
fig. 1-7.

radiculis anastomosantibus instructis, ramis spicatis (plerumque) integerrimis approximatis stricte erectis, cellulis filorum primariorum hyalinis cytioplasmate contracto (in statu vegeto ?), diametro transversali (inferiorum) duplo—triplo longioribus et æqualibus (superiorum), cellulis ramorum tumidis, omnibus in sporangia zoogonidia evolventia transmutatis diametro æqualibus et dimidio brevioribus. Diam. cellularum filorum primar. 0,0112—0,013 mm. $(\frac{1}{180}$—$\frac{1}{138}'''$ Engl.) Diam. ramorum sporidiferorum 0,0084—0,0112 mm. $(\frac{1}{283}$—$\frac{1}{180}'''$ Engl.) Altitudo plantulæ 1—1,5 mm.

In *Nitellæ* cellulis et in foliis muscorum.

Hoc *Stigeoclonium elegantissimum* cum *S. debili* et *uniformi*, Kütz. (Tab. Phyc. iii. tab. 3), aliqua similitudine consentit. Primum differt ramulis longioribus longius distantibus non fasciculato-racemosis ; secundum, ramificatione verticilliformi caulis primarii. Hæc species tres cum *S. gracili*, Kütz., subspecies formant specieis unæ typicæ.

Tab.V., Fig. i.—1, *Nitellæ* pars cum cæspitulo *Stigeoclonii* insidente $(\frac{6}{1}^0)$;— 2, fili singuli summa pars major aucta, omnes cellulæ zoogonidia procreantes $(\frac{240}{1})$.

2. **Stigeoclonium subtile**, *Reinsch in Journ. Linn. Soc.* xv. 217 ; minutissimum, parasiticum, ex filis sterilescentibus tenuioribus longioribus integerrimis erectis e filis procumbentibus dense intertextis crassioribus ortis formatum, cellulis ramulorum erectorum tenuioribus diametro 4plo-8plo longioribus, cellulis filorum procumbentium latioribus diametro subæqualibus, filis propagativis paulo crassioribus, cellulis zoogonidia procreantibus cellulis filorum sterilescentium multo brevioribus subquadraticis arctissime conjunctis. Diam. filiorum erectorum 0,0048—0,0056 mm. $(\frac{1}{567}$—$\frac{1}{318}'''$ Engl.)

In muscorum foliis, in *Nitellæ* et *Vaucheriæ* cellulis, et in *Schizosiphontis kerguelensis* trichomatibus.

Hoc *Stigeoclonium* ramulis prolongatis tenuissimis flagelliformibus erectis ex ramulis crassioribus ortis aliqua similitudine consentit cum *S. setigero*, Kütz. (Tab. Phyc. iii. tab. 5), quod distinguitur cæspitulis multo majoribus fluctuantibus (usque tres lineas longis).

Cæspituli tantum fila propagativa procreantes haberi possunt pro Speciem propriam. In singulis speciminibus plantulæ inveniuntur et fila sterilescentia et fila propagativa. Ulteriora paulo crassiora sed breviora saepe inveniuntur ex uno ramulo evoluta cum filis sterilescentibus. Cæspitulos quoque singulos in *Hormosiphontis* sp. n. physcumate crescentes una cum *Choreoclonii procumbentis* gen. n. cæspitulis observavi ; in his plantularum duarum infimæ partes adeo sunt inter se coalitæ ut plantulas duas valde diversas in cohæsione genetica putare possis.

Choreoclonium, genus novum.[*] Plantula parasitica ex filis ramosis procumbentibus densius aut laxius intricatis substrato dense adpressis interdum parenchymatice inter se conjunctis formata ; cellulæ rectangulares usque quadraticæ.

_____

[*] κορέω expando, κλώνος cluuis.

Propagatio ?—Synon. Genus s. n. in Reinsch, Contribut., p. 76, tab. iv. (Chloroph.) descriptum et delineatum genus ad *Chætophoraceas* spectans, *Stigeoclonio proximum.*

1. **Choreoclonium procumbens**, *Reinsch in Journ. Linn. Soc.* xv. 217. Cellular. diam. 0,0028—0,0041 mm. ($_{\gamma 1}^{1}$ — $_{3 6}^{1}$''' Engl.) Cellular. longit. 0,0112—0,0224 mm ($_{1 1 9}^{1}$—$_{0 4}^{1}$''' Engl.)

In foliis muscorum et in *Nitellæ* cellulis.

Hanc plantulam primo observavi anno 1872 in Germania in plantis aquaticis (*Hottonia, Utricularia*) crescentem, deinde in compluribus formis variis locis Germaniæ. In contributionibus meis formas varias in uno genere conjunctas sine nomine recepi; post plantulam Kergulenensem inventam nimirum dubitare possum in identitate plantularum e locis duobus remotissimis.

Tab. IV., Fig. ix.—-1, folii musci aquatici pars cum plantula singula minore in nervo folii crescenti ($^{A 6}_{1}$);—2, alteri folii pars cum plantula majore obtecta ($^{3 6}_{1}$).

1. **Draparnaldia subtilis**, *Reinsch in Journ. Linn. Soc.* xv. 218; filis ramisque primariis hyaliuis, ramis e basi repetito dichotome ramosissimis, ramulis furcatis acutis plerumque in pilum hyalinum ex cellulis compluribus exstitutum longe productis, cellulis infimis fili primarii diametro æqualibus cytiodermate crasso lamelloso, cytioplasmate subbomogeneo subtilissime granulato, cellulis superioribus diametro usque duplo longioribus, cellulis ramulorum diametro usque triplo longioribus, cytioplasmate dense granulose. Diam. fili primarii 0,0168—0.0232 mm. ($_{1 2 2}^{1}$—$_{8 4}^{1}$'''.) Diam. ramulorum 0,0056—0,0084 mm. ($_{3 1 8}^{1}$—$_{2 4 3}^{1}$''' Engl.) Plantulæ altitudo 1—2 mm.

In *Vaucheriæ* cellulis et in muscis aquaticis in plantulis dispersis radiculis numerosis radicantibus. Hæc plantula elegans tantummodo in speciminibus paucis observata differt a ceteris *Draparnaldiis* et magnitudine et loco natali.

2. **Draparnaldia distans**, *Kütz., Tab. Phyc.* iii., tab. 14, fig. 2; forma tenuis, cellulis fili primarii duplo-quadruplo diametro transversali longioribus, ramis primariis perpaucis, ramulis sparsis crebrioribus brevioribus cum ramulis longioribus in ambitu lanceolatis perpaucis intermixtis, ramulis ultimis plerumque in pilum achroum cellulare attenuatis, cellulis ramulorum tumidis diametro subæqualibus. Diam. cellular. fili prim. 0,0278—0,056 mm. ($_{1 5}^{1}$—$_{3 8}^{1}$''' Engl.)

In speciminibus exsiccatis cum *Zygnemate* intermixtis. "In a freshwater pool on the W. of Swain's Bay." (DISTRIB. Europa.)

1. **Proterderma viride**, *Kützing.* Familiæ singulæ in foliis musci aquatici laxius insidentes, in magnitudine cellularum ac forma (0,0084 mm.; $_{2 8 3}^{1}$''' Engl. diam.) cum speciminibus Franconicis maxime consentiunt.

1. **Zygogonium torulosi**, *Kütz., Tab. Phyc.*, tab. 14, fig. 1; forma crassior. Cellulæ diametri transversalis dimidio brevioribus (ante divisionem usque æqualibus) cytiodermate interiore crassissimo plurilamelloso, cytiodermate exteriore

subtoruloso.   Diam. cellular. 0,0168—0,0196 mm. $(_1\frac{1}{2\sigma}—_{\Gamma}\frac{1}{10}'''$ Engl.)   Diam.
filorum (c. indum.) 0,033—0,0393 mm. $(_\frac{1}{03}—_{3}\frac{1}{4}'''$ Engl.)

In cœspitulis inter *Hormosiphon coriaceum*, var.   " In moist places near Vulcan Cove."—DISTRIB. *Z. torulosi* Europa orientalis.

In filis singulis observantur sicut in speciminibus Europæis cellulæ subsphæricæ laterales filis adhærentes indumento crassissimo velatæ.   Quæ cellullæ—nullo modo cellulæ propagativæ—oriuntur in hoc *Zygogonio* et in *Z. anomalo* divisione longitudinali interdum incidente cellularum singularum fili.—Hæc forma a forma typica in Tab. Phycol. deliueata cellulis angustioribus cytiodermate crassiore et filis crassioribus distinguitur. *Z. torulosum*, *Kütz.*, cum serie specierum : *Z. ericctorum*, *anomalum*, *delicalulum*, a Cl. Rabenhorst (Fl. Eur. Alg. ii., p. 254) in una specie contrahuntur, sed characteres constanter observatæ horum *Zygogoniorum* a speciebus Kützingianis discedere mihi non permiserunt.

2. **Zygogonium tenuissimum,** *Reinsch in Journ. Linn. Soc.* xv. 218; filis tenuissimis cellulis diametro duplo longioribus (et paulo minus) regulariter rectangularibus, cytiodermate subcrasso homogeneo hyaliuo, cytioplasmate contracto colore luteo-viridi granulis majoribus instructo.   Diam. cellular. 0,0068—0,0084mm. $(_{..}\frac{1}{40}—_{2}\frac{1}{43}'''$ Engl.)

In singulis filis inter *Scytonemom castaneum* dispersis, " near Vulcan Cove."—Differt a *Z. delicalulo* et *Z. salino* cellulis longioribus, a *Z. gracili* et *Z. Ralpsii* cellulis brevioribus, ab omnibus *Zygogoniis* autem filis multo tenuioribus.

1. **Spirogyra longata,** *Kütz.*, *Tab. Phyc.* v. tab. 20, fig. 1 ; cellular. diam. 0,039—0,05mm. $(_{5}\frac{1}{4}—_{4}\frac{1}{2}'''$ Engl.)   Longitudo cellularum 5plum—7plum latitudiuis.

In a freshwater pool W. of Swain's Bay (specim. exsiccat.).

Structura fasciæ spiralis latæ anfractibus 4is—5is maxime consentit cum speciminibus Europæis. Fila omnia incopulata sunt latiora (usque duplo) filis formæ communis Europeæ per totam Europam diffusæ.—(DISTRIB. Europa, America borealis.)

2. **Spirogyra** *Spec.* ; Cellularum diam. 0,0278—0,0393mm. $(_{7}\frac{1}{3}—_{3}\frac{1}{4}'''$ Engl.) Longitudo 4plum—5plum latitudinis. Fila omnia incopulata ad quandam speciem *Spirogyræ* spectantia, quæ pertinet ad *Spirogyras* cytiodermate in utroque polo cellulæ nec protenso nec replicato ; sed structura fasciarum spiralium adeo est trausmutata ut numerus fasciarum et forma certe non potest explicari.

Fila dispersa inter *Draparnaldiam distantem*.   " Iu a freshwater pool on the W. of Swain's Bay" (specim. exsiccat.).

1. **Sirogonium sticticum,** *Kütz.*   Cellular. diam. 0,045—0,050mm. $(_{4}\frac{1}{7}—_{4}\frac{1}{2}'''$ Engl.)   Fasciæ *chlorophyllaceæ* ternæ—quaternæ in quaque cellula nucleis ex substantia proteinicis cum jodinis agentia fuscescentibus majoribus (nunc decoloratis) instructæ. Qui uuclei sunt majores nucleis speciminum ex Germania.

In singulis filis incopulatis inter *Zygnemæ* cæspites. "In a freshwater pool W. of Swain's Bay."—(DISTRIB. Europa borealis et centralis.)

1. **Zygnema Vaucheri,** *Agardh; Z.* SUBTILE, *Kütz., Tab. Phyc.* v. tab. 16, fig. 1. Diam. cellular. 0,0168 mm. ($\frac{1}{120}'''$ Engl.) Fila omnia incopulata. "In a stream W. of Swain's Bay, 20. 1. 1875."—(DISTRIB. Europa.)

Speciminum in charta siccatorum cytioplasmatis structura distincte non perspicua, attamen tinctura intensive lutea in charta effusa post aqua conspersa *Zygnemam* certissime indicat.

2. **Zygnema affine,** *Kütz., Tab. Phyc.* v. tab. 16, fig. 3.—Diam. cellular. 0,0196—0,0224mm. ($\frac{1}{610}-\frac{1}{94}'''$ Engl.) Longitudo duplum usque triplum latitudinis. Structura cytioplasmatis in singulis cellulis bene conspicua.

In filis dispersis singulis inter *Draparnaldiam distantem.* "In a freshwater pool W. of Swain's Bay."—(DISTRIB. Europa orientalis.)

1. **Bulbochæte** *Species.* Cellular. diam. med. 0,013mm. ($\frac{1}{151}'''$ Engl.) Cellularum longitudo triplum latitudinis, cellula basalis oblongo-pyriformis pedicello pediformi breviore.

In foliis musci aquatici.

A *Bulbochæteis* genere species observavi tres, specimina omnia sine florescentia et fructificatione. Apud species singulas cellulæ filorum et dimensionibus et forma consentiunt; qua de causa difficillime possunt determinari specimina sterilescentia. Cellulæ basalis et cellularum forma ac magnitudo aliqua similitudine consentit cum *B. crenulata*, Pringsh.

2. **Bulbochæte** *Species.* Cellula basalis pyriformis basi attenuata. Diam. cellular. 0,013 mm. ($\frac{1}{151}'''$ Engl.)—Forsan species propria.

Plantula singula in musci aquatici folio crescens.

3. **Bulbochæte** *Species.* Cellular. diam. 0,0168—0,0224mm. ($\frac{1}{120}-\frac{1}{94}'''$ Engl.) Cellulæ paulo longiores quam latiores. Cellula basalis late pyriformis basi subito in pedicellum pediforme augustata; ramificatio repetito dichotome ramosa.

In foliis musci aquatici, in *Nitella* cellulis, et in *Schizosiphonte kerguelensi.*

Aliqua similitudine consentit cum specie nova descripta et delineata in contributionibus meis, p. 81, tab. xiv. (chloroph.), fig. 4.

Specimina fere omnia observata ostendunt incrementi intensi, ut observamus in speciminibus collectis in mensibus Maii et Junii in Europæ regionibus mediis. In evolutione setarum, quæ brevissimo tempore fit, cellulæ matricalis cytiodermatis partes per cellulam filialam se evolventem ad latus premuntur, et brevi tempore permanentes in hoc statu, post evolutis setis, dejiciuntur. Quæ lamellæ cytiodermatis in hoc tempore vitæ in plurimis plantulis sunt observandæ. Quod factum tempus certum in historia incrementi hujus plantulæ nobis indicat. Plantulæ Kerguelenenses in hieme collectæ repræsentant statum evolutiones ejusdem spatii in nostris regionibus mensibus Maii et Junii respondentem. Idem factum est observatum in speciminibus *Nitellæ antareticæ*, quæ sunt in primo statu florescentiæ.

**   L

1. **Oedogonium delicatulum**, *Kützing.* Diam. cellular. 0,0056mm. ($\frac{1}{318}'''$ Engl.), longitudo 3plum—4plum latitudinis.

In singulis filis sterilescentibus in foliis musci aquatici, et in *Nitella* insidentibus observatum; cellulæ basalis forma filorum latitudo, cellularum longitudo cum speciminibus Europæis maxime consentit.—(DISTRIB. Europa, frequens.)

2. **Oedogonium** *Species.* Diam. cellular. 0,0112mm. ($\frac{1}{160}'''$ Engl.), longitudo 8plum—9plum latitudinis. Non possunt determinari specimina propter penuriam florescentiæ et fructificationis. Complures species *Pedogoniorum* maximo consentiunt in cellularum forma ac magnitudine, ex qua causa non potest dici de earum fila sterilia aliquid certi.

In filis singulis sterilescentibus in foliis musci aquatici.

1. **Coleochæte scutata**, *Brébisson;* Phyllactidium setigerum, *Kützing, Tab.* *Phyc.* iv., tab. 87, iii.

In foliis musci aquatici in familiis planis planitieformibus plus minusve regulariter circumscriptis forma. Familiæ complures in foliis angustioribus et in foliis quibus lamina deest, in massas irregulariter sphæricas aggregatæ.

Phyllactidium setigerum, Kütz., et *Coleochæte scutata*, Kütz., plantulæ synonymicæ sunt, prima haberi potest forma setigera *Coleochæleis scutalæ*. Speciminum Kerguelenousium setæ paulo sunt longiores et robustiores setis specimiuum Europeorum. (DISTRIB. Europa, America borealis.)

2. **Coleochæte irregularis**, *Prings., Jahrb. f. wissensch. Bot.* 1860, ii., tab. i., fig. 6, tab. vi. fig. 3–9. Diam. cellular. 0,0112—0,0126mm. ($\frac{1}{150}—\frac{1}{148}'''$ Engl.)

In *Nitellæ* superficiem cellularum dense incrustans.—DISTRIB. Europa, orientalis.

Cum speciminibus Europæis maxime consentit plantula Kerguelenousis in magnitudine et forma cellularum. Specimina plantulæ quæ observavi e diversis locis Europæ (Galliæ et Germaniæ) consentiunt quoque in loco natali (in *Nitellæ* cellulis*, plerumque in *Nitella synearpa*), et in crescendi modo.

1. **Aphanochæte repens**, *Al. Braun.* Diam. cellular. 0,011mm. ($\frac{1}{150}'''$ Engl.)

In foliis musci aquatici. Sine *oogoniis*, fila in foliis sæpe inter Algas minores (*Leptothrix*, *Tolypothrix*) dispersa. Cum speciminibus Europæis in omnibus partibus consentit.—(DISTRIB. Europa, America borealis.)

Extra formam typicam filis substrato dense adpressis invenitur quoque fòrma peculiaris in speciminibus singulis dispersis forma. Filis in corpuscula sphærica et uvæformia accumulatis (pressione in filis singulis soluta).

1. **Gongrosira pachyderma**, *Reinsch in Journ. Linn. Soc.* xv. 218. Aquatica, parasitica, pulvinulos ex cellulis absque ordine (rarius in modo parenchymatis) cohærentibus formans, cellulis irregulariter sphæricis usque ellipsoidicis (omnibus? fructiferis) cytioplasmate granuloso colore pallide luteo-viridi, cytiodermate crassissimo plurilamelloso (usque cellulæ diametri transversalis dimidio

æquali). Diam. cellularum (cytioderm. incl.) 0,039—0,051mm. ($\frac{1}{8'1}$—$\frac{1}{42}$''' Engl.) Cytiodermatis crassitudo 0,0041—0,0056mm. ($\frac{1}{517}$—$\frac{1}{817}$''' Engl.) In foliis musci aquatici. *G. De Baryana*, Rabenh. (Fl. Europ. Alg. ii., p. 388; Alg. Eur. Nro. 223), proxima species (in lapidibus lignisque pr. Francofurtem ad Oderam, Itzigsohn et De Bary.), differt cellulis minoribus in filis subramosis dispositis, cytiodermate tenuiore.

MELANOPHYCEÆ ET RHODOPHYCEÆ.

Rhizocladia,* *Noc. Genus* (ad *Phæosporeas* Thuret spectans, *Pleurocladiæ* Al. Braun proximum).

Plantula ex strato procumbente ex filis ramosis substrato viventi dense adhærentibus fornato et ex filis erectis ramosis fructiferis exstituta. Cellulæ filorum procumbentium primo rectangulares, ætate provectiore ovales usque lageniformes. Fila erecta singula aut bina ex cellulis filorum procumbentium orta, primo integra et ex cellulis æqualibus formata, demum subramosa et fructifera et ex cellulis inæqualibus formata. Trichosporangia in apice filorum erectorum ex 3is—5is cellulis quadraticis usque rectaugularibus formata. Oosporangia?

1. **Rhizocladia repens,** *Reinsch in Journ. Linn. Soc.* xv. 220.   Character idem generis.   Longitudo cellul. explic. filor. repent. 0,0097—0,0112mm. ($\frac{1}{210}$—$\frac{1}{189}$''' Engl.)   Latitudo cellular. filor. erect. juvenil. 0,0041mm. ($\frac{1}{517}$''' Engl.) Latit. trichosporang. 0,0058mm. ($\frac{1}{312}$''' Engl.)   Plantulæ explicitæ altitudo 0,056—0,089mm. ($\frac{1}{31}$—$\frac{1}{213}$''' Engl.)

In foliis muscorum aquaticorum et in cellulis *Nitellæ.*

Hanc plantulam primo habui proximam algæ jam descriptæ et delineatæ in contributionibus meis (p. 76, tab. 14, Chlorophylloph.), quæ planta est constituta ex filis ramosis procumbentibus. Ad hanc plantam nunc ad *Chætophoraceas* positam pertinentes formas postea inveni et ex filis procumbentibus et ex ramulis brevissimis erectis exstitutas. Sed per compages peculiares ex cellulis compluribus brevioribus supra positis exstitutas sine dubio trichosporangiis *Melanophycearum* proximas hujus plantulæ positionem veram agnoscimus in systemate. Cellulæ singulæ dispersæ observantur in filis erectis singulis, ceteris cellulis crassiores et breviores (oosporangia?), quarum natura vera in statu vivente tantummodo agnosci potest. A genere *Pleurocladia* unico aquæ dulcis hucusque cognito *Melanophycearum* organis peculiaribus fœcundationis præditarum differt incrementi modo ac filorum structura.

Tab. V., Fig. ii.—1, plantulæ juvenilis filis sterilescentibus æqualiter altis particulus, cellulæ filorum repentium arctissimæ intricatæ ($\frac{120}{1}$); fig. 2, aliud specimen cæspituli expliciti filis erectis in statu vario evolutionis, filum singulum in apice trichosporangium (*a*) evolvens ($\frac{120}{1}$).

* μίζα radix, κλάδος ramus.

** M

1. **Batrachospermum minutissimum,** *Reinsch in Journ. Linn. Soc.*
xv. 220; e minimis, oculis inarmatis vix conspicuum, parasiticum, filis integerrimis
erectis singulis aut perpaucis aggregatis, articulis inferioribus subcuneiformibus
apice paulo incrassatis, cellulis corticalibus 4is—6is obtectis superioribus nudis
rectangularibus, ramulis verticillorum integerrimis (rarius singulis ramulis instructis) æqualibus, apicibus paulo angustatis ex cellulis 5is—7is exstitutis apicem
fili versus sensim decrescentibus, inferioribus articulorum longitudine subæquantibus,
summis duplo-triplo longioribus, cellulis ramulorum rectangularibus usque subquadraticis, cytiodermate exteriore tenuissimo vix conspicuo, cytioplasmate subhomogeneo colore obscure olivaceo-viridi; fructificatio?—Diam. articulorum
0,0041—0,0056mm. ($\frac{1}{567}$—$\frac{1}{318}$''' Engl.)   Diam. ramulorum 0,0041mm. ($\frac{1}{567}$''' Engl.)
Plantulæ altitudo 0,37—0,45mm. ($\frac{1}{4}$—$\frac{1}{3}$''' Engl.)

In *Nitella* cellulis et in muscorum foliis, in filis singulis sparsis cum aliis
algis (*Tolypothrix*, *Leptothrix*) intermixtis, rarius in cæspitulis parvulis.

B. tumidum, in *Chara vulgari* crescens (Reinsch, Contributiones, p. 60, tab. xliv.,
Rhodoph. fig. 1–5) a speciebus hucusque cognitis proxima species, sed valde
diversa dimensionibus omnium partium multo majoribus (3–4 lineas longa), verticillorum ramulis numerosis repetito dichotome ramosis; in ramulorum cellularum
forma aliqua similitudine consentit.

[The fresh-water species recorded by Reinsch are 106, to these may be added the
following mentioned in the Antarctic Flora:—*Oscillatoria purpurea,* Hook. f.& Harv.,
Calothrix olivacea, H. f. & H., *Ulva cristata,* H. f. & H., *Mastodia tessellata,*
H. f. & H., *Trypothallus anastomosans,* H. f. & H., *Nostoc commune,* Vaucher, and
N. microscopicum, Carm., making a total of 113 species. This interesting Antarctic
island, therefore, so far as explored, appears to be very rich in certain forms of freshwater Algæ.—*G. Dickie.*]

VII.—*Fungi.*

By the Rev. M. J. BERKELEY, M.A., F.L.S.

[The Fungi collected in Kerguelen Island amount to 9 or 10 (the tenth being still an undetermined form).*

Dr. Hooker obtained 2 species in the winter (May and June) 1840; Mr. Moseley 3 in addition to the same, during summer (December and January) 1873-4; Mr. Eaton, also in summer, 5 determinable species, and 1 that could not be identified (*see* footnote), besides the species found by Dr. Hooker.

Until a few days before Midsummer (*i. e.* Christmas) no Fungi were seen in the vicinage of the English Observatory Bay. The first to appear was the common mushroom, a single specimen of which was found on an island in the sound by some officers from H.M.S. "Volage." Later in the summer the other four species came up in a few places on the mainland. They were not by any means of frequent occurrence, and probably scarcely any of them would be found at the time of year corresponding with the date of Dr. Hooker's visit to the island.—*A. E. Eaton.*]

1. **Agaricus** (GALERA) **kerguelensis**, *Berk. in Journ. Bot.* v. 51 (1876); *et in Journ. Linn. Soc.* xv. 22. Cæspitosus, fulvus, pileo e breviter campanulato convexo lævi carnuloso, margine tenui striatâ, stipite æquali apice pulverulento-granulato, lamellis distantibus ventricosis adnatis.

Amongst moss in a bog on the eastward portion of the base of a promontory E. of Vulcan Cove, January 1875, *Eaton.*

Cæspitose, attached by abundant mycelium. Pileus ½ inch across; stem nearly 1 inch high, ¼ to ¾ line thick; principal gills about 12 in number, shortly but truly adnate, and not in the least decurrent.

It is far more fleshy than any variety of *A. hypnorum*, to which species no doubt it is closely allied; and while agreeing with *A. embolus* in possessing comparatively few gills, it differs from that species in the mode of their attachment.

2. **Agaricus** (GALERA) **hypnorum**, *Batsch.; Berk. in Journ. Linn. Soc.* xv. 53.

Hab.—On *Azorella.* January 1874. *Moseley.*

Spores ·0004 inch long.

3. **Agaricus** (NAUCORIA) **furfuraceus**, *Pers.; Berk. in Journ. Linn. Soc.* xv. 221.

---

* This species is referred to by Mr. Eaton (in Proc. Roy. Soc. 1875, May. xxiii. 355) as "a peculiar "parasite on *Azorella*, which grows out of the rosettes" of the leaves "in the form of a clear jelly, which "becomes changed into a firm yellowish substance of indefinite form." It was common on the sides of hills in the neighbourhood of the observatory towards the end of December, and a series of examples was preserved in spirit, but they could not be worked out.

In the same bog as *A. kerguelensis*, and at the same time, *Eaton*.

4. **Agaricus** (Naucoria) **glebarum,** *Berk. in Flor. Antaret.* 417, t. clxii.
fig. iii.; *et in Journ. Linn. Soc.* xv. 53.

On *Azoretta*, January 1874, Kerguelen Island, *Hooker, Eaton*. (Marion Island,
Moseley. On tufts of *Botax*, Falkland Islands, *Hooker*.)

Spores ·0003 inch long.

5. **Agaricus** (Psalliota) **campestris,** *Linn.; Berk. in Journ. Linn. Soc.*
xv. 221. A. (P.) arvensis, *Eaton in Proc. Roy. Soc.* xxiii. 355.

On an island near Observatory Bay, in Royal Sound, 16th December 1874. A
solitary specimen, *Eaton*. (Almost cosmopolitan.)

1. **Coprinus atramentarius,** *Fries; Eaton in Proc. Roy. Soc.* xxiii. 355;
et in Journ. Linn. Soc. xv. 222 (footnote). .

Two or three specimens were found singly close to the margins of two of the
lakes among the hills near Observatory Bay, in February 1875, *Eaton*.

2. **Coprinus tomentosus,** *Fries; Berk. in Journ. Linn. Soc.* xv. 53.

On dung, January 1874, *Moseley*.

1. **Peziza** (Sarcoscyphæ) **kerguelensis,** *Berk. in Flor. Antaret.* 451,
t. clxiv. fig. iii.; *Cooke, Myeot.* fig. 134; *et in Journ. Linn. Soc.* xv. 53, 222.

Christmas Harbour, May and June, on bare boggy ground near the sea, growing
amongst *Confervæ, Hooker*. Amongst dwarfed *Leptinetta plumosa*, on wet ground
close to the shore, growing in rings, rare. One ring on an island in Swain's
Bay, January, and another on the mainland near Observatory Bay, February 1875,
Eaton. Royal Sound and Betsy Cove, *Moseley*. (Hermite Island, Cape Horn, alt.
1,000 ft., *Hooker*.)

1. **Sphæria herbarum,** *Pers.*

On dead stems of *Pringlea, Eaton*.

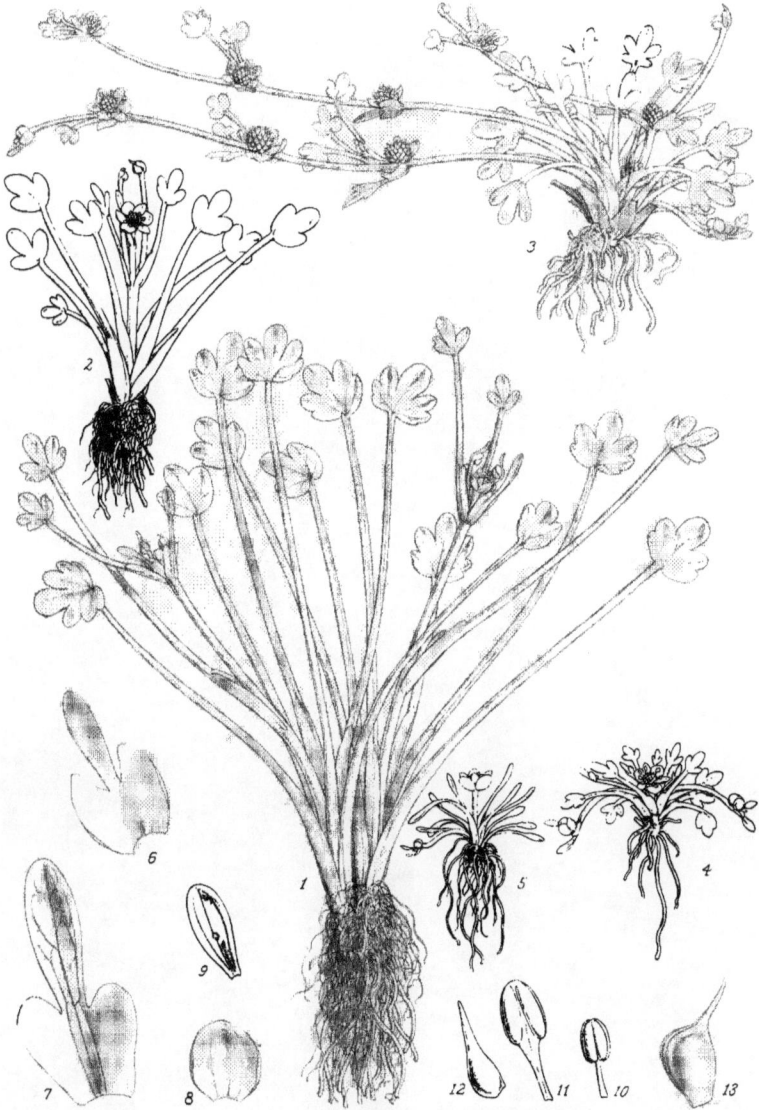

J.N.Fitch lith.

RANUNCULUS TRULLIFOLIUS. *Hook.f.*

Fitch imp

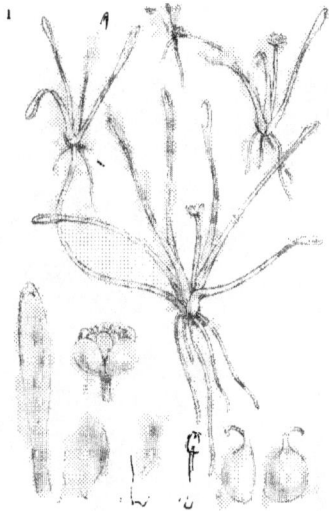

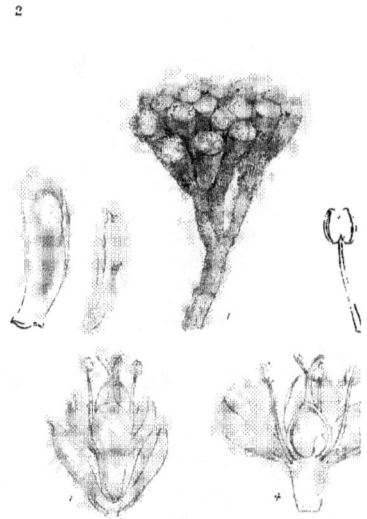

Ranunculus Meselev: *Hrt*

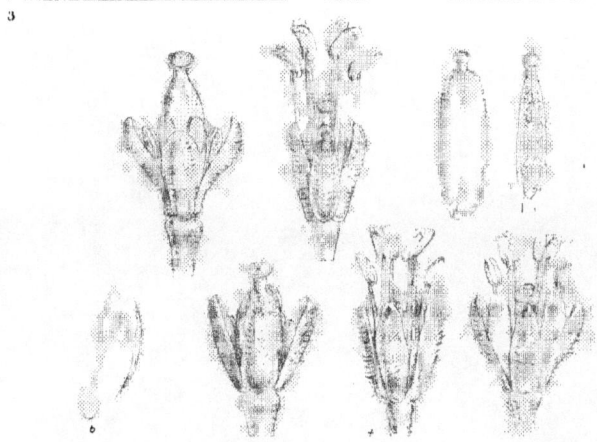

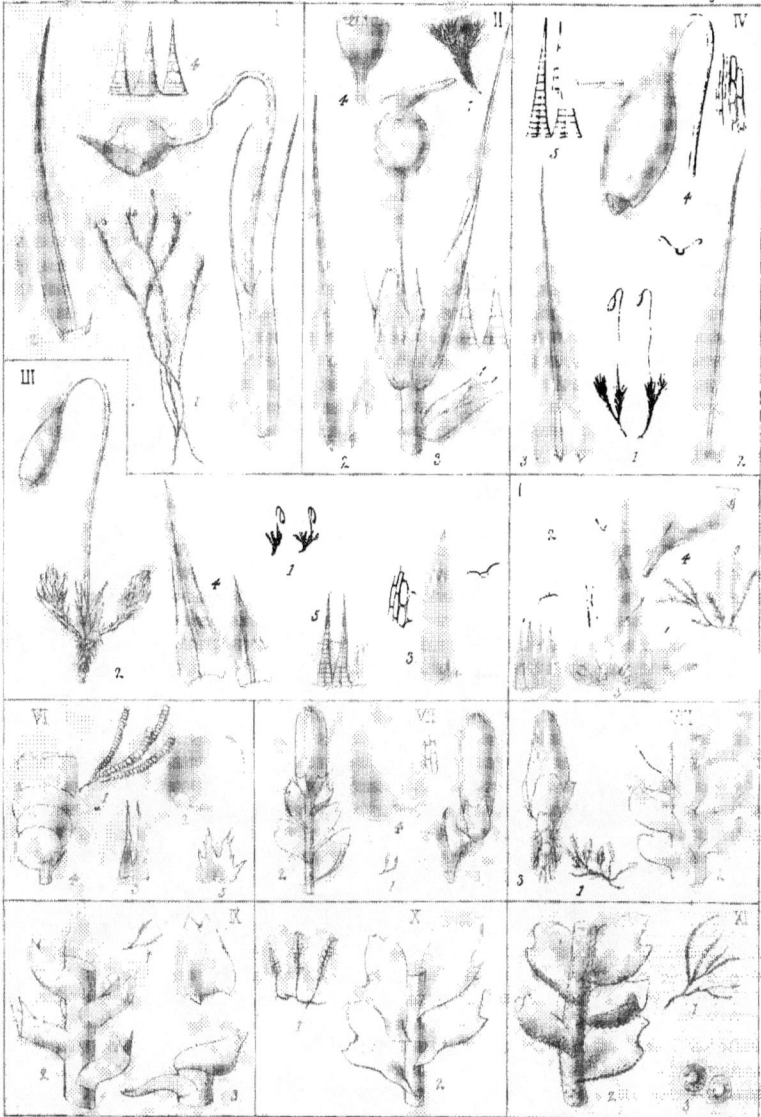

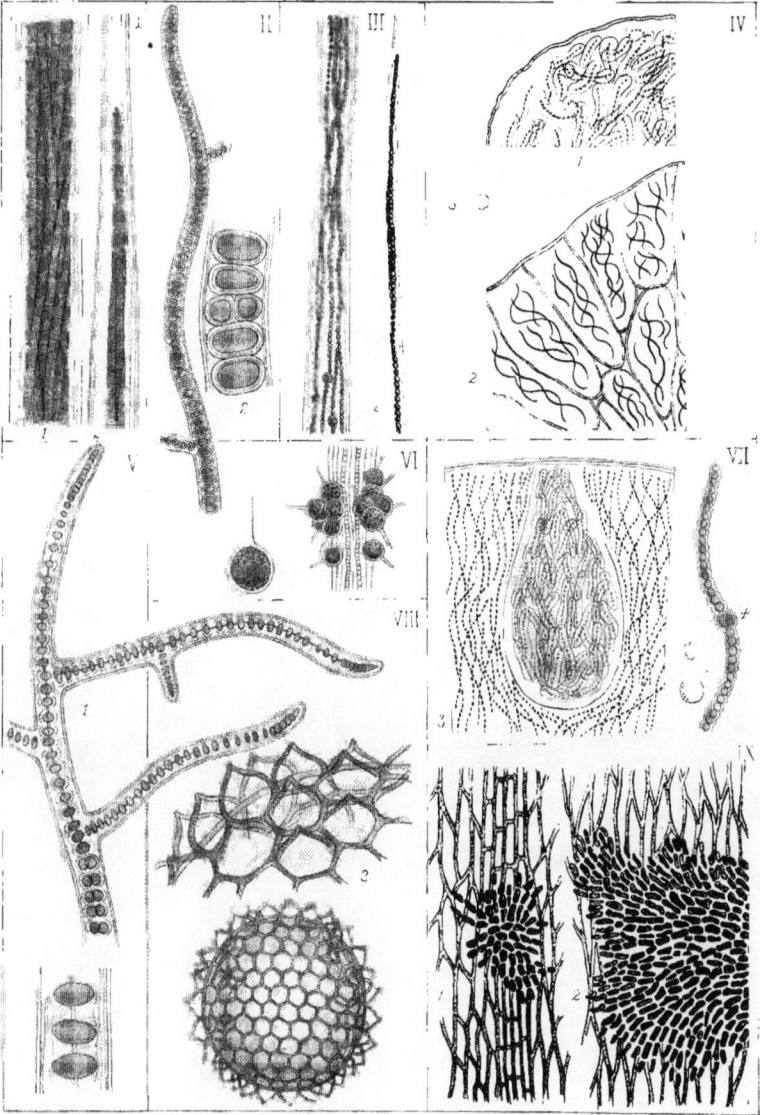

I

II

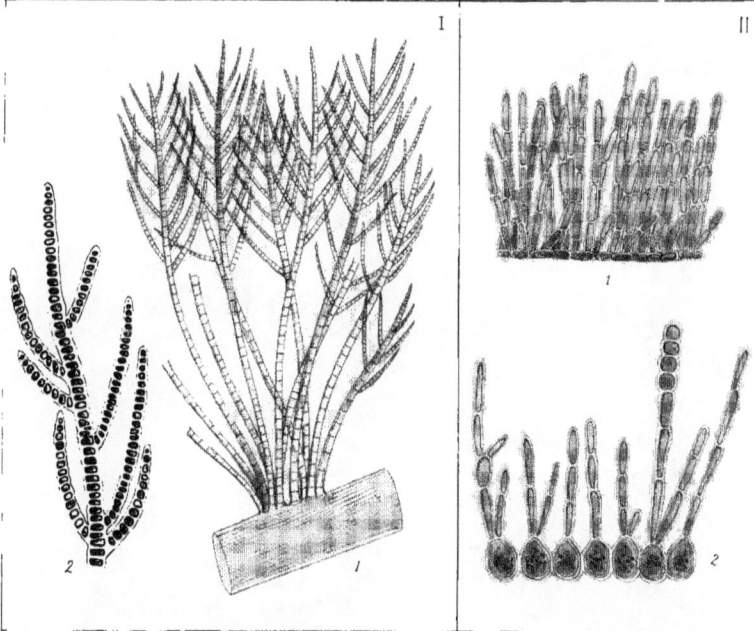

III

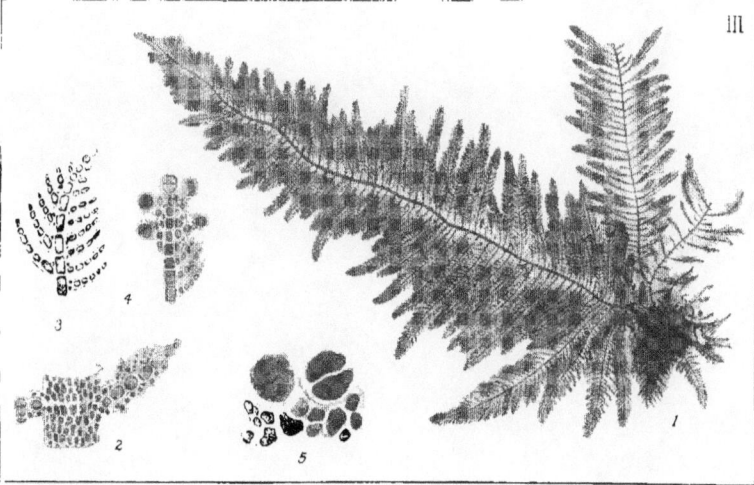